Aug 22, 2012
Bloomington, IN.

Victor McKusick and the History of Medical Genetics

Krishna R. Dronamraju • Clair A. Francomano
Editors

Victor McKusick and the History of Medical Genetics

Springer

Editors
Krishna R. Dronamraju
Foundation for Genetic Research
Houston, TX, USA

Clair A. Francomano
Greater Baltimore Medical Center
Harvey Institute of Human Genetics
Baltimore, MD, USA

ISBN 978-1-4614-1676-0 e-ISBN 978-1-4614-1677-7 (eBook)
DOI 10.1007/978-1-4614-1677-7
Springer New York Heidelberg Dordrecht London

Library of Congress Control Number: 2012938639

Printed on acid-free paper

Springer is part of Springer Science+Business Media (www.springer.com)

Foreword

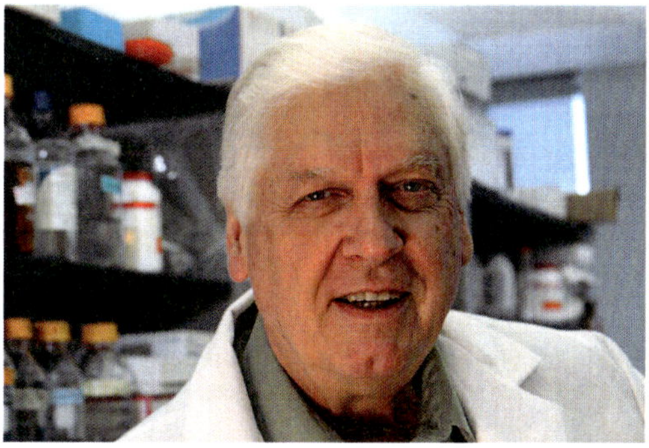

Hamilton O. Smith received an A.B. degree in mathematics at the University of California, Berkeley in 1952 and the M.D. degree from Johns Hopkins University in 1956. After 6 years of clinical work in medicine (1956–1962), he carried out research on Salmonella phage P22 lysogeny at the University of Michigan, Ann Arbor (1962–1967). In 1967, he joined the Microbiology Department at Johns Hopkins. In 1968, he discovered the first TypeII restriction enzyme (HindII) and determined the sequence of its cleavage site. In, 1978 he was a co-recipient (with D, Nathans and W. Arber) of the Nobel Prize in Medicine for this discovery. Subsequently, he studied DNA methylases and nucleases in *Haemophilus influenzae* Rd and discovered this organism's sequence-specific DNA uptake during genetic transformation. In 1994–1995 he collaborated with J. Craig Venter at The Institute for Genomic Research (TIGR) to sequence *H. influenzae* by whole genome shotgun sequencing and assembly. In July 1998, he joined Celera Genomics Corporation where he participated in the sequencing of the *Drosophila* and human genomes. In November 2002, he left Celera to join the new Institute for Biological Energy Alternatives formed by J. Craig Venter. In 2005, this Institute merged with TIGR to form the J. Craig Venter Institute where he is currently leading the synthetic biology group in an effort to make a synthetic bacterial cell.

I first met Victor McKusick in 1954 when I was a medical student at Hopkins. Among my many teachers, he stands out in my mind because he was young, enthusiastic, and high tech. I remember particularly one of his lectures in which he described the latest diagnostic tools being developed and used in cardiology, including if my memory serves me correctly, electro phonocardiograms. This application of new technologies to medical diagnosis was exciting to me. It was through his lectures that I developed a deep interest in cardiology which led to my subsequently taking electives in EKG interpretation and a one-quarter course with Helen Taussig on auscultation in the diagnosis of congenital and valvular heart disease.

It was not until the 1960s, when I began my research career in genetics and molecular biology that I learned of Victor's rise as an authority on human genetic diseases. He had become world famous for his encyclopedic compilation of all known human genes and genetic disorders, published as a series of volumes under the title, *Mendelian Inheritance in Man (MIM)*. The database was religiously kept up to date as new diseases were discovered. When he spoke at meetings or seminars, he would proudly show slides displaying the ever-growing number of volumes of MIM. When the human genome project started in earnest around 1990, I recall attending a dinner party with Victor and we began speculating about the number of human genes. We decided to make a bet and the winner would get a free dinner. I guessed 100,000 genes, but Victor thought there would be much fewer, around 50,000. It turned out that we were both wrong, but he was much closer. The real number was about 25,000. Unfortunately, I never had the opportunity to pay off the bet.

It was because of Victor's preeminence in medical genetics that our paths crossed again when I joined with Craig Venter at Celera Genomics in 1998 to sequence the human genome. Victor was a strong supporter of our efforts, and I saw him frequently at meetings of the Celera Scientific Advisory Board, and also later as a member of the Venter Institute Board of Trustees in Rockville, MD. Victor was a gentle human being who lent his vast knowledge, wisdom, good judgment, and prestige to our meetings. We were all very saddened by his passing.

Hamilton O. Smith, M.D.

Acknowledgements

I express my gratitude to Dr. Anne McKusick for her kind cooperation in putting this book together. In addition to contributing a fine chapter, Anne generously provided much help in obtaining additional material and photographs. I am also grateful to Judge Vincent McKusick for his fine contribution about the McKusick twins, as well as helping with the identification of family photographs from their childhood. I am appreciative of the Foreword provided by Hamilton Smith, who knew Victor well as a student and later as a colleague.

I am appreciative of all the contributors, especially Kenneth and Victor McKusick, for kindly permitting me to reproduce their eulogies. Dr. Clair A. Francomano not only contributed two finely written chapters and a eulogy but also provided much assistance in preparing the book for publication. I thank Melanie Tucker at Springer for her cooperation and assistance in publishing this book.

Houston, TX, USA Krishna R. Dronamraju

Editor Biographies

Krishna R. Dronamraju is President of the Foundation for Genetic Research, Houston, and a Visiting Professor of the University of Paris. He was a student and close associate of J.B.S. Haldane, receiving his Ph.D. in human genetics from the Indian Statistical Institute, and later worked with Dr. Victor McKusick at the Johns Hopkins University School of Medicine. Dr. Dronamraju is the author of 20 books and over 200 papers in genetics and biotechnology. He was an Advisor to President Bill Clinton's administration and was a member of the United States Presidential delegation to India in 2000. He served on the Recombinant DNA Advisory Committee of the U.S. National Institutes of Health, Washington, D.C.

Clair A. Francomano attended Yale College as an undergraduate and received her M.D. from Johns Hopkins University School of Medicine. She worked with Dr. McKusick as a student, doing research among the Lancaster County Amish, and trained in Internal Medicine and Medical Genetics at Johns Hopkins from 1980 to 1984. She joined the full-time Hopkins faculty in 1984. In 1994 she became Chief of the Medical Genetics Branch at the National Human Genome Research Institute, National Institutes of Health, where she served as Clinical Director from 1996 to 2001. From 2001 to 2005 she was Chief of the Human Genetics and Integrative Medicine Section in the Laboratory of Genetics, National Institute on Aging. She joined the Greater Baltimore Medical Center faculty in 2005 as Director of Adult Genetics at the Harvey Institute of Human Genetics. Her research interests over the years have centered on Hereditary Disorders of Connective Tissue and Skeletal Dysplasias, areas to which she was introduced by Victor McKusick in the early days of her career.

Contents

Contributors

Digamber S. Borgaonkar Laboratory of Neurogenetics, National Institute of Aging, National Institutes of Health, Bethesda, MD 20892, USA

11 Stabler Circle, Wilmington, DE 19807, USA

Elof Axel Carlson Stony Brook University, Stony Brook, NY, USA

Institute for Advanced Study, Indiana University, PO Box 8638, Bloomington, IN 47407, USA

Aravinda Chakravarti Center for Complex Disease Genomics, McKusick-Nathans Institute of Genetic Medicine, Johns Hopkins University School of Medicine, 733 N. Broadway, BRB Suite 579, Baltimore, MD 21205, USA

Krishna R. Dronamraju Foundation for Genetic Research, P.O. Box 27701–0, Houston, TX 77227, USA

Malcolm A. Ferguson-Smith Department of Veterinary Medicine, Cambridge Resource Centre for Comparative Genomics, Cambridge University, Madingley Road, Cambridge, CB3 0ES, UK

Clair A. Francomano Harvey Institute for Human Genetics, Greater Baltimore Medical Center, 6701 North Charles Street, Baltimore, MD 21204, USA

Peter S. Harper Institute of Medical Genetics, Cardiff University School of Medicine, Heath Park, Cardiff, CF14 4XN, UK

Edison T. Liu The Jackson Laboratory, Bar Harbor, ME 04609, USA

Don M. Long Department of Neurosurgery, Johns Hopkins Hospital, Baltimore, MD 21204-3741, USA

Anne Bishop McKusick Department of Medicine, Johns Hopkins University School of Medicine, Baltimore, MD 21205, USA

Vincent L. McKusick Pierce Atwood LLP, Merrill's Wharf, Portland, ME 04101, USA

Nicholas Avrion Mitchison University College London, 13 Framfield Road, London M5 1UU, UK

UCL Research Department of Immunology, Cruciform Building, 90 Gower Street, London WC1E 6BT, UK

Reed E. Pyeritz Departments of Medicine and Genetics, Center for the Integration of Genetic Healthcare Technologies, The Perelman School of Medicine at the University of Pennsylvania, Philadelphia, PA 19104-4283, USA

Hospital of the University of Pennsylvania, Penn Tower 1115, 3400 Spruce St, Philadelphia, PA 19104, USA

David L. Rimoin[*] Medical Genetics Institute, Cedars-Sinai Medical Center, 8700 Beverly Blvd., Los Angeles, CA 90048, USA

Giovanni Romeo University of Bologna Medical School and European Genetics Federation, via S. Petronio Vecchio, Bologna 42–40125, Italy

Hamilton O. Smith J. Craig Venter Institute, 10355 Science Center Drive, San Diego, CA 92121, USA

David J. Weatherall Weatherall Institute of Molecular Medicine, John Radcliffe Hospital, University of Oxford, Headington, Oxford OX3 9DS, UK

---

[*]Unfortunately, Dr. Rimoin passed away while the book is in press

Chapter 1
Victor McKusick

Krishna R. Dronamraju

Victor receiving the National Medal of Science, June 13, 2002

K.R. Dronamraju (✉)
Foundation for Genetic Research, P.O. Box 27701-0, Houston, TX 77227, USA
e-mail: kdronamraj@aol.com

K.R. Dronamraju and C.A. Francomano (eds.), *Victor McKusick and the History of Medical Genetics*, DOI 10.1007/978-1-4614-1677-7_1,

Introduction

I am grateful to Victor's twin, Vincent, for providing information about family history and background. Victor was the first physician/scientist in the family.

The McKusick twins were born and raised in Parkman, Maine, which stands almost exactly at the geographical center of the state, some 50-plus highway miles northwest of Bangor. Their ancestral line has been identified with Parkman since 1840. Their father was Carroll L. McKusick (1882–1969), and their mother was Ethel May Buzzell (1882–1944). Both parents were in the teaching profession and encouraged their children's education. In their direct ancestral line, the twins had no forebear who attended college before their father Carroll went to Bates College in the year 1900. All received a basic education in Parkman one-room schoolhouses and little beyond. Yet all the twins' direct forebears are remembered as outstanding citizens, active in the public affairs of their community, including support of local education.

In an autobiographical note written on the occasion of the Lasker Award, Victor recalled that he was first exposed to science (and to genetics) when he went up to Tufts to take premed studies in 1940. Victor and Vincent split up when they went to college, Vincent to Bates and Victor to Tufts, partly to avoid competing with each other for scholarship support but also because of diverging career goals.

Victor's desire to take up medicine was the result of a 10-week stay at the Massachusetts General Hospital, at the age of 15, when he developed an abscess of the left axilla and a superficial spreading ulcer of the right elbow that would not heal. Eventually, he was treated with sulfanilamide which had become available the year before, and the lesions healed and stayed healed. Victor later wrote that during that process, he saw much of medicine and decided that was the field for him.

Johns Hopkins

In 2006, Victor described himself as an "autodidact" in genetics, a quality he shared with many gifted scientists. He first came to know genetics in his last semester at Tufts when he took an elective course in genetics with Professor Paul A. Warren, "an inspiring teacher, who made the topic of classic Mendelism exciting." He finished up his third academic year at Tufts by January 1943. He would have continued there except for the fact that, in the fall of 1942, he learned that Johns Hopkins Medical School was accepting applications from undergraduate students who had not yet received a college degree, for an entering class in March 1943. This was a sudden shift for Johns Hopkins, brought on by the wartime exigencies, waiving their usual requirement of a bachelor's degree that had been strictly followed since its opening in 1893. Furthermore, he was strongly attracted to Johns Hopkins since a cover story appeared in *Time* magazine in January 1939, which focused on the Hopkins Professor of the History of Medicine, Henry Sigerist, and his views on socialized medicine. It also contained an account of the early history of Johns Hopkins and of its Institute of the History of Medicine. That was the only medical school application he had ever made and was accepted to matriculate in March 1943.

Victor completed four academic years of medical school in three calendar years, graduating in March 1946. He continued at Hopkins with an internship in internal medicine and fully expected, after some postdoctoral training, to return to Maine to enter the general practice of medicine. However, his ambitions for the internship on the prestigious Osler Medical Service proved stronger, which led him to academic medicine.

Medical Genetics

Victor's interest in medical genetics began in June 1947, when a teenager named Harold Parker became his patient near the end of his Osler internship. That was his introduction to the polyps-and-spots syndrome. In the next 2 years, four additional cases of this kind came to his attention. Three of them were members of the same family, indicating an autosomal dominant inheritance. When he found out that Harold Jeghers in Boston also had five cases, Victor collaborated with him in writing up these ten cases in two papers in successive issues of the *New England Journal of Medicine* [1]. This syndrome, consisting of jejunal and other intestinal polyps and melanin spots of the lips, buccal mucosa, and digits, was later called the Peutz-Jeghers syndrome, in a paper from the Mayo clinic in 1954 [2]. However, Victor pointed out that Jonathan Hutchinson in London first described the pigmentary changes in identical twins in 1896 [3].

Pleiotropism

For the genetic interpretation of the polyps-and-spots syndrome, Victor sought the advice of Bentley Glass, who was then a member of the Biology Faculty of Johns Hopkins University. It was Glass who impressed upon Victor that pleiotropism of a single mutant gene was a more likely explanation for the association between polyps and spots rather than genetic linkage.

Victor's early work in cardiology is well known. Early in his career, his training in cardiology included reading electrocardiograms in the Johns Hopkins Heart Section under the direction of Elliott V. Newman. Later, from 1948 to 1950, he worked in a cardiovascular unit at the US Marine Hospital in Baltimore under Luther L. Terry, who subsequently became famous as the US Surgeon General who blamed cigarette smoking for lung cancer and other ailments. Victor was involved in several other studies in cardiology, especially of heart sounds and murmurs by the method of sound spectrography, which had been developed at the Bell Telephone Laboratory for analyzing speech sound. These studies resulted in a monograph entitled *Cardiovascular Sound in Health and Disease* [4].

Victor was greatly interested in historical studies in genetics; his writings include the initial chapter of successive editions of Emery and Rimoin's *Principles and Practice of Medical Genetics* [5], biographical articles on Jonathan Hutchinson, Frederick Parkes Weber, Walter Stanborough Sutton, and Marcella O'Grady Boveri. Victor was especially eloquent on the contributions of the London multispecialist, Jonathan Hutchinson, dedicating two volumes of *Clinical Delineation of Birth Defects* to the memory of Hutchinson.

Marfan Syndrome

While still on the junior faculty of Johns Hopkins, Victor pursued a detailed study of Marfan syndrome as well as four others, grouped under the category (I) "heritable disorders of connective tissue": Ehlers-Danlos syndrome, osteogenesis imperfecta, pseudoxanthoma elasticum, and Hurler syndrome (the prototype of the mucopolysaccharidoses). This work was ultimately published as a book entitled *Heritable Disorders of Connective Tissue* [6]. Victor was greatly impressed by the multiple organ systems that were affected by the pleiotropic action of the mutation involved. He focused attention on all the Marfan patients he could, fully realizing that Johns Hopkins was a "superb site" for such studies because of its excellent departments of ophthalmology, pediatric and adult cardiology, orthopedics, and other specialties. Victor preferred to say "the Marfan syndrome" (rather than Marfan syndrome) because it makes it clear that the surname is merely a tag. He wrote, "After all, Marfan described the skeletal features only" [7, p. 29].

Because of the autodidactic nature of his genetic knowledge, Victor found it helpful to associate himself, as a junior faculty member at Hopkins, with other like-minded individuals which included Bentley Glass, Barton Childs, Abraham Lilienfield, and others, forming what was informally called the "Galton-Garrod Society."

Division of Medical Genetics

Medical genetics was formally institutionalized at Johns Hopkins on July 1, 1957, when Victor was invited by the Chairman of the Department of Medicine, A. McGehee Harvey, to take over the direction of a multifaceted chronic disease clinic that was initially developed by J. Earle Moore in 1952. Victor set out to develop a Division of Medical Genetics, within the Department of Medicine, based in that clinic which was renamed in honor of Dr. Moore. Victor justified this process by pointing out that genetic disease is the ultimate in chronic disease.

Victor later wrote that some of his colleagues were surprised when he switched from cardiology to genetics. There was a gradual transition. The Moore Clinic was earlier involved in a study of the genetics of heart disease and the initial support was provided by the National Heart Institute until the National Institute of General Medical Sciences was founded in the early 1960s. Victor continued to be involved in the Heart Sound Laboratory until 1962. Victor found that certain facilities of the old Moore Clinic, such as long-term follow-up mechanisms for studying the late manifestations of syphilis, and its close links to the Johns Hopkins School of Hygiene and Public Health, with its statistics and epidemiology departments, were ideally suited for the new medical genetics unit as well. Paradoxically, he also found that carrying out research, teaching, and patient care in a cramped space was ideally suited for fostering collaborations among the staff and fellows with multiple research interests in cytogenetics and biochemical aspects, as well as other disciplines in relation to genetic disease.

The Division of Medical Genetics was a beehive of activity. Victor's energy and enthusiasm were infectious. Many fellows came from Great Britain, which was mainly due to Victor's close friendship with Cyril Clarke in Liverpool. In his obituary of Cyril Clarke, Victor wrote, "The passing … of Cyril Clarke brought back memories of pleasant and productive interchanges between Liverpool and the Johns Hopkins Hospital over a 20 year period or more beginning in 1957. Cyril was a central figure in those transatlantic collaborations in training and clinical research …. We at Johns Hopkins are much indebted to Cyril for the able protégés he sent to Baltimore in the formative years of medical genetics here" [8]. Among the Liverpudlians, who came to the Moore Clinic, were David A. Price Evans, Peter Brunt, J. Michael Connor, Brian Hanley, Peter S. Harper, F. Michael Pope, Brian Walker, David Weatherall, J.C. Woodrow, and Ronald Finn (the latter worked with Julius R. Krevens). Malcolm Ferguson-Smith came down from Glasgow to establish the cytogenetics laboratory. Weatherall, Harper, and Ferguson-Smith are contributors to this book. These are listed in the chapter by Weatherall.

Victor was ably assisted by two colleagues in the Division of Medical Genetics; one was Edmond A. (Tony) Murphy in statistical genetics and Samuel H. ("Ned") Boyer in biochemical genetics. David Weatherall worked first with Ned Boyer in the Biochemical Genetics section of the Moore Clinic and later with C. Lockart Conley in the study of hemoglobinopathies and with others in the Department of Biophysics directed by Howard Dintzis.

Victor's Publications

Victor's major research contributions and their publications are described and discussed in this book by several colleagues. These include his work on Marfan syndrome and related disorders, his extensive work with the Amish populations [9], contributions to genetic nosology, his great interest in gene mapping, his massive compilation of Mendelian disorders and the online version (OMIM), his role in founding and guiding the Human Genome Organization (HUGO), his contribution to education and training in medical genetics through a series of summer courses in Bar Harbor (USA), Bologna (Italy), and Beijing (China), and his most valuable studies in the history of science and medicine.

Victor was a prolific writer (see Table 1.1). During his lifetime, he published an enormous output of 772 research papers, books, reviews, and miscellaneous reports, which are listed in the Appendix. During the years 1949–1973, he published 538.

Table 1.1 Publications of Victor McKusick

Years	Number of all publications
1949–53	37
1954–58	112
1959–63	125
1964–68	122
1969–73	142
1974–78	74
1979–83	38
1984–88	32
1989–93	25
1994–98	29
1999–03	23
2004–08	13
Total	772

He published fewer (234) but still a large number in the remaining years 1974–2008. Once he became chairman of the department of medicine in 1973, there was a sharp drop in the number of publications. For example, in the five preceding years (1969–73), his publications numbered 142, as compared to 74 during the years 1974–78. This is understandable because of additional administrative work and other responsibilities which he assumed in 1973. Also, the work on the catalogs of Mendelian Inheritance in Man, Human Gene Mapping, and HUGO took up more and more time in later years. He was publishing until the very end; during the last 2 years, 2007–2008, there were nine publications, altogether, an impressive record of lifetime performance.

A scientist's penetrating insights and perceptions are not always found in his major works but in reviews of the works of others, in brief reports, and in correspondence, as well as personal conversations. Victor is no exception.

I start with a minor publication, an early review [10], of *Human Heredity* by the French biologist, Jean Rostand, (translated from French) which was published in 1961. Rostand was the son of Edmond Rostand, creator of *Cyrano de Bergerac*. Victor's review opens with the sentence "For a small book, this one packs a powerful lot of misinformation and misconception." He points out that the responsibility rests with the author, translator, and the Philosophical Library whose bad judgment it was to produce an English edition. It was first published in 1952, and the book translated in 1961 did not take into account the advances that took place during the intervening years. Victor contrasted this book with Penrose's "beautiful little" *Outline of Human Genetics* [11]. Rostand's book falls short of its avowed objective: "to introduce the greatest number of people into the sovereign dignity of knowledge," whereas Penrose's little book fulfilled that purpose "quite satisfactorily." Victor goes on to make other penetrating (and amusing) comments: "Men with many daughters will be intuitively suspicious of the view repeated on page 126 that 'there is probably a correlation between the tendency to produce boys and the virility of the father.'" "And the so-called simple inheritance of rare hereditary disease traits is misleading because it misses the important concept of heterogeneity of entities which phenotypically appear to be homogeneous."

Health of the Presidents

One amusing book, which caught Victor's attention because of his historic interest, was "The Health of the Presidents" by Rudolph Marx, a Los Angeles surgeon. Victor's comment, "His book undoubtedly makes entertaining light reading" [12]. Almost all presidents seemed to have had serious psychologic problems which led to other health problems: two examples are Taft's overeating and obesity, and the alcoholism of Franklin Pearce, as well as many others. Victor commented that "such collections of information about the health and illnesses of famous persons serve a useful mnemonic purpose in clinical teaching and practice: Cerebral thrombosis may occur in the 40s with survival for as many as 20 years thereafter, as in the case of Mrs. William Howard Taft." Another comment about a well-known scientist, Louis Pasteur had a stroke at 46, yet made his greatest contributions in the following 27 years.

Victor took the opportunity to make several corrections while making medical and historic observations of his own. Virginia was *not* the southernmost of the original 13 colonies. The reviewer was not aware that alcohol can be incriminated as an allergen in causing chronic bronchitis. "Habituation to whiskey" is meant, not "habitation to whisky," although the latter expression has interesting imagery. About Taft, the author wrote that he "developed the 'effort syndrome' – pain in the chest, shortness of breath, and heart consciousness after physical exertion, typical for angina pectoris." Victor's comment, "It was true angina pectoris from which Taft suffered. 'The effort syndrome,' which indeed has the mentioned symptoms as features, is a synonym for neurocirculatory asthenia (also called Da Costa's syndrome, cardiac neurosis, soldier's heart, etc.). Marx makes no mention of Taft's auricular fibrillation." Victor's conclusion, "Both the author's history and his medicine are thin in many places. However, he has produced a book that tells an absorbing series of stories without dispensing too much or too serious misinformation."

Misconceptions in Human Genetics

Writing in 1971 [13], Victor found it necessary to list 14 genetic misconceptions. His article has a timeless quality. This was extremely useful then (and still is today) as we come across these false ideas. Certain fundamental truths need to be repeated often.

1. "Congenital is synonymous with genetic." Congenital merely means present at birth. Exceptions are found in both directions. Some genetic disorders are not congenital in the usual sense, and many congenital malformations do not have a predominantly genetic cause. For instance, Huntington's chorea is genetic but not congenital, whereas rubella embryopathy is congenital but not genetic.
2. "If a disorder is inherited, a chromosome analysis will show abnormality." Most Mendelian disorders have no chromosome abnormality which can be demonstrated with existing techniques.

3. "A buccal smear provides full information on the chromosomes." The buccal smear only tells the maximal number of X chromosomes per cell.
4. "If a genetic disorder is dominant, all children of an affected person will be affected; conversely, if all children of normal parents are affected by a genetic disorder, this is evidence of dominant inheritance." This indicates ignorance of Mendel's laws and the meaning of "dominance."
5. "When individuals of only one sex are affected in a family, it indicates sex-linkage of the disorder." Once again, this indicates ignorance of basic principles of genetic transmission.
6. "A disorder that occurs in multiple siblings with normal parents is not hereditary." Recessive inheritance is characterized by affected sibs with normal parents, and a recessive disorder is as genuinely hereditary as a dominant one.
7. "Consanguinity brings out sex-linked disorders, for instance, hemophilia was frequent in the inbred royal families of Europe." Consanguinity increases the occurrence of homozygous affected females but has no effect on the frequency of affected males.
8. "The occurrence of a hereditary syndrome composed of two or more manifestations is the result of close linkage on the same chromosome of separate genes, each resulting in one of the individual manifestations." All "mendelizing" syndromes studied in full detail to date have been found to have their basis in the pleiotropic effect of a single mutant gene. Linkage produces no permanent association of traits because even closely linked genes become separated through the process of crossing-over.
9. "Dominantly inherited disorders tend to increase in severity with transmission from generation to generation, a process called anticipation. For example, age of onset, an expression of severity, tends to be lower in affected children than in their affected parents." Ascertainment bias is responsible for apparent anticipation. On the average, only in milder cases of many dominant disorders, the offspring are expected to be more severely affected than their parents.
10. "Dominant disorders are common; recessive disorders are rare." There is no correlation between the frequency of a disorder and its mode of inheritance.
11. "Inbreeding causes a build-up of 'bad genes' in populations." Inbreeding does not directly change the gene frequencies. It does change genotype frequency; it increases the frequency of homozygotes. If the homozygote is at a disadvantage, inbreeding actually results in a decrease in the deleterious genes.
12. "Dominant disorders are more severe than recessive disorders." As a general rule, just the opposite is the case; however, severity may be associated with either type.
13. "If a couple has had three children born with a given recessive disorder, the chance that the fourth child will also be affected is vanishingly small." The risk of an affected child from two carrier parents remains 1 in 4, regardless of previous family history.
14. "Genetic disease is not treatable": False. Many genetic diseases are treatable.

Genetic Nosology: On Lumpers and Splitters

Victor believed that certain fundamental facts are worth repeating. He emphasized the concepts of pleiotropism and genetic heterogeneity in his now classic paper, "On lumpers and splitters, or the nosology of genetic disease," which appeared in the *Birth Defects, Original Article Series*, in 1969 [7]. This paper is noted for delineating several fundamental problems, for instance, what constitutes a "genetic entity," what is the impact of inbreeding on nosology, and what are the problems in naming genetic entities. He mentioned examples from his own work, pleiotropism as a leading concern with respect to Marfan syndrome, whereas genetic heterogeneity became increasingly the focus in studies of the genetic mucopolysaccharidoses and the separation of homocystinuria from the Marfan syndrome. This became evident in successive editions of *Heritable Disorders of Connective Tissue* [6].

Victor compared the classification of disease with the taxonomy of plants and animals. Taxonomists like nosologists can be either lumpers or splitters. But there is one important difference. The nosologist's major concern is whether syndromes A and B are one and the same entity or distinct entities. The taxonomist, on the other hand, constructs a branching classification based on the phylogeny, the components in his classification bearing varying degrees of genetic relationship to each other, based on their descent from a common ancestor.

The fundamental questions tackled by Victor include the following: What constitutes a genetic entity? How does one identify genetic heterogeneity (including clinical, genetic, and biochemical methods)? Practical difficulties include rarity of the phenotype in the population, and small size of the families, and the naming of genetic entities. The phenotype resulting primarily from a specific and unitary factor is an entity. He wrote that delineation of genetic entities is on safe ground if a fundamental biochemical defect or a specific chromosomal anomaly is identified.

Inbreeding in the Amish

Victor discussed nosology in relation to inbreeding, for instance, the delineation of "new" genetic entities with recessive inheritance is enhanced in inbred communities. It was, in fact, his interest in inbreeding in the Amish populations which brought us together because of my research on inbreeding in India with J.B.S. Haldane [14–18]. In his genetic studies of the Amish, Victor combined multiple research interests. He wrote, "in the 1960s, skeletal dysplasias became an area of both clinical and research nosologic interest because of their relationship to the heritable disorders of connective tissue, because of studies of dwarfism in the Amish … which we initiated in 1963, and because of collaboration with Little People of America, Inc., which began in 1965 when I first attended the annual national convention of this fraternal organization of persons of short stature. Skeletal dysplasias became a major interest of many who were my fellows in that period including David Rimoin, Judith G. Hall, and Charles Scott, who like me became honorary life members of Little People of America." [19].

Victor's expectation that "new" recessive disorders would be found in the Amish population was "richly fulfilled" [16]. The first phenotype which was studied in detail in the Amish was dwarfism, two forms of which were found in the Amish of Lancaster County, Pennsylvania. One was the Ellis-van Creveld (EvC) syndrome or six-fingered dwarfism. The second form of recessive dwarfism, a previously unknown entity, was designated cartilage-hair hypoplasia (CHH), later called metaphyseal chondrodysplasia, McKusick type. A large number of other "new" disorders in the Amish were described by several investigators including the Hopkins group. The resulting publications on the Amish were compiled and edited by Victor in a convenient volume published by the Johns Hopkins University Press in 1978.

From 1968 to 1972, Victor arranged 1-week-long conferences each year, entitled "Clinical Delineation of Birth Defects" which were held at the Johns Hopkins Hospital and were supported by the National Foundation – March of Dimes (MOD). They were centered around genetic nosology and covered all areas of medicine. The conferences and the resulting publications enabled colleagues from all departments of the Johns Hopkins Hospital as well as elsewhere an opportunity to participate and share their experiences with rare or not so rare genetic disorders.

Gene Mapping

Victor once wrote, "I am not certain why, in the late 1950s, I became enthralled with mapping genes on human chromosomes" (19). One of his first attempts at gene mapping was in collaboration with Ian Porter which involved the map distance between the loci for G6PD deficiency and color blindness on the X chromosome. Linkage studies in the Moore Clinic continued in the 1960s in collaboration with James Renwick who (with Jane Schulze and David Bolling) wrote one of the first computer programs for linkage analysis. It was used in the analysis of possible linkage of XG and CB. Renwick was also involved in several other Moore Clinic linkage studies, including assignment of the Duffy blood group locus to chromosome 1 in 1968, first assignment of a specific locus to a specific autosome. Other linkage studies from the Moore Clinic included mapping of the interval between nail-patella syndrome and the adenylate kinase locus.

The HGM (Human Gene Mapping) workshops were a forerunner of the Human Genome Organization (HUGO). The first HGM workshop was planned in New Haven in 1973 by Frank Ruddle, but the funding from the March of Dimes Foundation (MOD) was organized by Victor. These workshops served an important function in the compilation of gene map information, both published and unpublished, over a period of two decades. HGM3 was organized by Victor in Baltimore in 1975.

Victor played a key role in convincing the National Foundation that birth defects fall into the domain of medical genetics. He sent a copy of the first edition of his *Heritable Disorders of Connective Tissue* to Thomas M. Rivers, who was Vice President of Research at the March of Dimes (MOD). As Victor recalled later, Rivers was a loyal Hopkins medical alumnus, with training in pediatrics at the Johns

Hopkins Hospital. Shortly afterward, Victor was invited to join the MOD medical advisory committee in 1959. The MOD was the sole funder of the Bar Harbor Short Course for its first 25 sessions and continued to provide some support afterward.

The Human Genome Project

A major contribution of Victor was his proposal at the International Conference on Birth Defects sponsored by the MOD in The Hague in 1969, that mapping all the genes would be a useful approach to understanding the basic derangements in birth defects. He stated later that his proposal reflected the exuberant mind-set that followed the first moon landing. In 1980, Victor proposed that it should be possible to map all the genes by the year 2000. Walter Gilbert pointed out that complete sequencing would be necessary just to find all the genes.

Genomics

At the Cold Spring Harbor symposium in May 1986, which was entitled "The Human Genome," Victor presented a paper on the status of the human gene map [20]. He thought it was an "eye-opener" to molecular biologists. It was about that time that Victor was approached by Brian Crawford of Academic Press who invited Victor to edit a new journal on mapping and sequencing. Victor was at first reluctant to undertake the new task but was encouraged to do so when Frank Ruddle agreed to join him as coeditor in chief. They were about to call the new journal *Genome,* but it was already preempted by the *Canadian Journal of Genetics and Cytology* which had renamed their journal *Genome*. Victor later recalled that a long evening session of the editorial board was held, over beer, in July 1986, and Thomas Roderick of the Jackson Laboratory in Bar Harbor, Maine, came up with the name *Genomics* as the title.

Founder President of HUGO

From the beginning, HUGO was supported by funds from the Howard Hughes Medical Institute (HHMI), which also supported the development of OMIM beginning in 1985 and for several years afterward.[1] HHMI, in fact, played the same

[1] Sir Walter Bodmer, who succeeded Victor as president of HUGO, said of him, "Victor was an outstanding Medical Geneticist, indeed a role model for many. He had an enormous influence on the field, including his strong support for the Human Genome project and was the first President of the Human Genome Organisation (HUGO). Victor's passing away is really the end of an era. He had such an enormous influence on the development of medical genetics and was a friend and mentor to so many people. He and his wife Anne were most approachable and friendly people, and in Victor McKusick's death, I feel I have lost a real friend."

supporting role as had the March of Dimes with reference to the Bar Harbor courses, the conferences on the Clinical Delineation of Birth Defects, and most of the Human Gene Mapping workshops.

HUGO was founded in 1988 as a coordinating agency for the global effort. It was proposed by Sydney Brenner, James Watson, Lee Hood, and others at the Cold Spring Harbor Symposium, and Victor was invited to serve as organizing president (Victor's Hugo)! Victor later reminisced that he was invited because of his active role in the HGM workshops and the ongoing record he had maintained of the status of the human gene map. The first task was to establish a founding council to write a mission statement and bylaws for HUGO. The first meeting of the council was held at Montreux, Switzerland, at the suggestion of James Watson, and HUGO was incorporated in Geneva. Victor noted that there were five nobelists at that meeting, including Watson, Dausset, Gilbert, Dulbecco, and Jacob, as well as several other scientists from 19 countries. It was incorporated in Delaware to facilitate handling of US funds.

The objectives of HUGO involved intellectual property rights (IPR), ethical issues, research materials, information sharing, and standardization of nomenclature. It was never conceived as a funding agency.

After the completion of his presidency of HUGO, Victor continued his work as cochair of the HUGO ethics committee and was a member of the ethics committee for the National Center for Human Genome Research (NCHGR). From the beginning, Jim Watson wisely decided that funding should be set aside in the budget of the NCHGR for the study of what he called ELSI, for ethical, legal, and societal issues.

The funding of the Human Genome Project by the federal government was mainly based on the NRC/NAS committee report on mapping and sequencing the human genome in 1988. That committee was chaired by Bruce Alberts. The report concluded that the project could be completed in 15 years at an annual budget of $200 million, and that "map first, sequence later" was the correct approach for several reasons. These are (a) mapping of DNA markers and cloned fragments would provide a useful scaffolding for sequencing, (b) sequence technology at that time needed further improvement, and (c) the sequence of genomes of other species should also be continued along with that of the human species, etc. The NIH project was initiated on October 1, 1990, with James Watson as the first director of the NCHGR. After his resignation, Francis Collins took over on January 1, 1993, and the center was given institute status within the NIH.

Celera Advisory Board

Victor was invited by Craig Venter to join the Celera Scientific Advisory Board in 1998 when he founded that company to undertake the complete sequencing of the human genome. Victor later recalled that his participation and association with

Hamilton ("Ham") Smith and Craig Venter was very stimulating. He thought that the "completion" of the Human Genome Project was appropriately accelerated by the competition that was introduced by the private effort. The motto of Celera, "Discovery cannot wait," added the right touch and excitement. Both the Venter map and the map created by the international consortium led by Francis Collins were announced at the White House on June 26, 2000.

Victor told me that he attended that ceremony and enjoyed it immensely. Later, I had learned all the details of that White House ceremony from my friend and neighbor in Houston, Neal Lane, who was then assistant to the president for Science and Technology and director of the White House Office of Science and Technology Policy under President Bill Clinton.

MIM and OMIM

Starting in the early 1960s, Victor compiled a catalog of human genes and genetic disorders. The basic idea of *Mendelian Inheritance in Man* [21] originated in the annual reviews of medical genetics which were undertaken by Victor and several associates during the years 1958–1963. With the onset of the genetic program in the Moore Clinic in 1957, a journal club was initiated, and its contributions became eventually a part of the MIM catalog, although the annual reviews of the literature were first published each year as 1 month's issue of the *Journal of Chronic Diseases*.

From the very beginning, the phenotypic nature of the mendelizing catalogs was emphasized. The subtitle of the first ten editions published was "Catalogs of Autosomal Dominant, Autosomal Recessive and X-linked phenotypes." Initially, the focus was on X-linked traits, and a monograph entitled "On the X Chromosome of Man" was published in the *Quarterly Review of* Biology in 1962 [22]. The second, the recessive catalog, was meant to be a resource in connection with study of recessive disorders in the Amish. The autosomal dominant catalog was then undertaken "for the sake of completeness" [19].

The first print edition of *Mendelian Inheritance in Man* (MIM) appeared in 1966, and it was maintained on the computer since 1964. For the 11th (1994) and 12th (1995) editions, the subtitle was changed to "Catalogs of Human Genes and Genetic Disorders," thus indicating the progress in the field. In the 1980s, MIM was prepared for online presentation, and a search engine enhanced its usefulness. Online access, OMIM, was first provided from the Welch Medical Library at Johns Hopkins beginning in 1987 and from the National Center for Biotechnology Information (NCBI) of the National Library of Medicine (NLM) from 1995 onward.

Victor once wrote that he reveled in the teaching of medical students and residents; "From the distinguished careers in genetics of several of those students and residents, it appears that through teaching in a framework of classic Oslerian medicine I may have helped bring genetics into mainstream medicine." [19].

References

1. Jeghers H, McKusick VA, Katz KH (1949) Generalized intestinal polyposis and melanin spots of the oral mucosa, lips and digits. N Engl J Med 211(993–1005):1031–1036
2. Bruwer A, Bargen JA, Kierland RR (1954) Surface pigmentation and generalized intestinal polyposis (Peutz-Jeghers syndrome). Mayo Clin Proc 29:168–171
3. Hutchinson J (1896) Pigmentation of lips and mouth. Arch Surg 7:290
4. McKusick VA (1958) Cardiovascular sound in health and disease. Williams and Wilkins, Baltimore
5. McKusick VA (2002) History of medical genetics. In: Rimoin DL, Connor JM, Pyeritz RE, Korf BR (eds) Emery-Rimoin principles and practice of medical genetics, 4th edn. Churchill Livingstone, Edinburgh, pp 3–36
6. McKusick VA (1956) Heritable disorders of connective tissue, 1st edn. Mosby, St. Louis
7. McKusick VA (1969) On lumpers and splitters, or the nosology of genetic disease. Birth Defects Orig Artic Ser 5:23–32
8. McKusick VA (2001) Persisting memories of Cyril Clarke in Baltimore. J Med Genet 38:284
9. McKusick VA (1978) Medical genetic studies of the Amish. Johns Hopkins University Press, Baltimore
10. McKusick VA (1961) Garbled information. Review of human heredity by Jean Rostand. Science 134:549–550
11. Penrose LS (1959) Outline of human genetics. Heinemann, London
12. McKusick VA (1962) Marx's health of the presidents. Review of the health of the presidents by Rudolph Marx. Bull Hist Med 36:383–384
13. McKusick VA (1971) Fourteen genetic misconceptions. Ann Intern Med 75:642–643
14. Bowman HS, McKusick VA, Dronamraju KR (1965) Pyruvate kinase deficient hemolytic anemia in an Amish isolate. Amer J Hum Genet 17:1–8
15. Dronamraju KR (1963) Genetic studies of the Andhra Pradesh population. In: Goldschmidt E (ed) The genetics of migrant and isolate populations, Williams and Wilkins, New York. Proceedings of an international conference on human population genetics, Hebrew University, Jerusalem, pp 154–159
16. Dronamraju KR (ed) (1968) Haldane and modern biology. Johns Hopkins University Press, Baltimore
17. Dronamraju KR (1989) Foundations of human genetics. Charles C. Thomas, Springfield
18. Dronamraju KR (ed) (1992) The history and development of human genetics: progress in different countries. World Scientific Publishing, Singapore/London
19. McKusick VA (2006) A 60-year tale of spots, maps, and genes. Annu Rev Genomics Hum Genet 7:1–27
20. McKusick VA (1986) The gene map of *Homo sapiens*: status and prospectus. Cold Spring Harb Sym 51:15–27
21. McKusick VA (1966) Mendelian inheritance in man: catalogs of autosomal dominant, autosomal recessive and X-linked phenotypes, 1st edn. Johns Hopkins University Press, Baltimore
22. McKusick VA (1962) On the X chromosome of man. Q Rev Biol 37:69–175

Chapter 2
Family Life and Victor's Meeting the Unexpected

Anne Bishop McKusick

Victor and I were married in June 1949 and, although impecunious, took a wonderful honeymoon in Paris, England, Scotland, and, owing to irregularities of our student flights, Geneva, Switzerland, as well. The old Transocean Airlines we traveled in were not in very good condition, and we lost one motor between Geneva and Shannon Airports, necessitating a 24-h layover in Shannon Airport. This gave us a chance to take a bus to Limerick for unexpected sight-seeing. Finally, we were on our way again and after a 19-h flight were back in the USA.

We settled into a North Broadway apartment on our return, and I completed my last year of medical school at Johns Hopkins, while Victor began a 2-year stint at the US Marine Hospital in Baltimore to fulfill his military obligation for deferment during World War II when he had attended medical school at Johns Hopkins. He worked in a cardiology group, while Luther Terry was the director of the hospital. At that time, Victor pointed out that electrokymography was useful in diagnosing constrictive pericarditis and experimented with the effect of lithium on the heart before he returned to Johns Hopkins as asst. resident, to become chief resident the following year, 1951–1952. He decided to put together a booklet of interesting cases he had seen during the year. By the time he had piles of page 1, page 2, and so on around a huge table, the house staff in a number equal to the number of piles ran around the table, picking up a sheet from each pile, thus collating pages of the book.

Our first car was bought in 1950, Victor having driven nothing but tractors previously. As he jackrabbited off in the morning, I prayed hard that I would see him again.

Living with my husband, Victor was always interesting. I realized early that he was strongly motivated to do something of significance. This meant never wasting a minute. It led to his early multitasking which may have begun when he watched his mother reading, knitting, and rocking at the same time or when, as a small boy on a farm in Maine, he had been told to "Mind the cows and take your book along."

A.B. McKusick (✉)
Department of Medicine, Johns Hopkins University School of Medicine,
Baltimore, MD 21205, USA
e-mail: ABMckusick@gmail.com

K.R. Dronamraju and C.A. Francomano (eds.), *Victor McKusick and the History of Medical Genetics*, DOI 10.1007/978-1-4614-1677-7_2,

Later, someone, impressed by his efficiency, joked that when Victor was stuck in an elevator, he emerged with two papers written.

Victor's intense interest in medical history came to my attention early when he tracked down the location of eight crates of drawings and paintings of unusual medical conditions seen by Sir Jonathan Hutchinson. Hutchinson had commissioned three artists to paint their clinical findings which he had encountered in his practice in England. Subsequently, Sir William Osler had ordered the crates sent to the Welch Medical Library at Johns Hopkins, and they had come to rest in its basement. Victor learned about them, and he and I sorted through them while he cataloged their contents. For several years, a picture of Sir Jonathan Hutchinson hung in our hall.

Our first child, Carol, was born in 1953. She was followed 10 years later by Kenneth and then 13 months after that by Victor Wayne.

Genetics Clinic

In 1957, Dr. A. McGehee Harvey, the chairman of medicine, asked Victor to head what had formerly been the venereal disease clinic of The Johns Hopkins Hospital and more recently, a chronic disease clinic. It had excellent capability for following the patients long term, and Victor was given permission to use it as a genetics clinic for the treatment of hereditary diseases. It was named The J. Earle Moore Clinic to honor Dr. Moore who had directed it formerly. Conferences each Thursday were exciting events at which patients were presented. They were attended by postdoctoral fellows studying under Victor and by physicians in other specialties concerned with the care of the patients. Still, other physician participants and observers came to these conferences from extramural sites. The clinic under Victor was interested in the different manifestations of given hereditary diseases and how they varied from family to family and within a family.

A journal club was set up, which met monthly in our home, and consisted of a review of genetics articles in the current medical literature. The articles were initially reviewed by one of the fellows for discussion by the group as a whole.

Fellows from other countries came to work in the Moore Clinic, and we learned things about their homelands and enjoyed the cosmopolitan group many of whom were, at first, from the UK. We were vitally interested in all of Victor's fellows and in their success in genetics.

Those around Victor sometimes learned that they too were part of his projects. I remember being called to the telephone at 10 a.m. in the midst of a busy arthritis clinic. It was Victor saying, with his endearing obliviousness, "How would it be if I brought 26 Geneticists home to dinner tonight?" Dr. Lionel Penrose, from the Galton Laboratory in London, was visiting, and Victor wanted his genetics fellows to meet him. For that sit-down dinner, all tables on the first floor were pressed into service. It was followed by Victor's inviting another course to come home for dinner, 150 strong. The guests chose which of 3 nights, during the week of the course, they would come. On that occasion, Dora, the housekeeper, joined the culinary force and picked 30 lbs. of crab in the course of her 3-day work week. The message that everyone worked was getting distributed.

Although things were busy and occasionally hectic, it was a life full of joys, excitement, and wonderful friendships.

One night, Dora, to be helpful, lit the fire Victor had laid in the living room fireplace. Unfortunately, she did not know the flu was closed. Twenty minutes before the guests arrived, the house was filled with smoke, so all hands turned to opening windows. Every year, for 5 years, that course came home for dinner. After 1 year, when Cecily, our Irish au pair girl, was leaving, she said to me, "Dr. McKusick, I'll never be scared of entertaining again."

Bar Harbor Course

In July 1959, following a visit to his birthplace and family home in Parkman Maine, Victor, our daughter Carol, and I were having lunch at Testa's Restaurant in Bar Harbor, ME. We had been joined by Dr. John Fuller of the Jackson Laboratory. It was then that the idea of a jointly sponsored genetics course occurred to Victor. It was discussed with Dr. Fuller and, either that night or the next, with Dr. Earl Green, director of the Jackson Laboratory, and his wife, Dr. Margaret Green, a mouse geneticist. There was agreement that the course should be tried.

It was planned that the Bar Harbor Course would draw one third of its faculty from The Jackson Laboratory, one third from Johns Hopkins, and the remainder from various institutions. The course was launched in 1960, and the students of the first year were for the most part department chairmen, wondering if genetics should be part of their curricula.

Lectures began at 8:30 a.m. and ran until 12:30 p.m. with a coffee break in the middle. Afternoons were at first free, for climbing mountain trails, and later, there were optional workshops after lunch. Individual dinners at 5 p.m. or 5:30 p.m. preceded the evening session, which resumed at 7:30 p.m., and ran until about 9 p.m. During the first year of the course, lectures were held in the ballroom of the Sir Harry Oakes mansion, which belonged to Bowdoin College and was rented by the course. Rudy and Connie Thayer and their daughter, Peggy, were doing the housework and getting the meals for the course of about 45 students. Professor Thayer was the coach of the Bowdoin debating team and a full professor in speech. We were like a large family with the Thayers joining us as hosts. A quiz given at the end of the first Bar Harbor Course asked, "Who murdered Sir Harry Oakes?"

The location of the Jackson Laboratory on Mt. Desert Island, adjacent to Acadia National Park, made it a particularly desirable site for students and their families to spend the 2 weeks of the course.

The trails on Mt. Desert Island included "The Beehive," a rather challenging climb. When our boys were about 11 and 12 and had a 12-year old friend with them, our family climbed "The Beehive," along with Dr. Paul and Sophie Crane, who had done it before, and their friends, the Crofoots. I realized on the way up that this ascent involved alternately holding and standing on U-shaped irons driven into a nearly vertical cliff. I found it very scary, but Victor and the boys were ahead of me, and I was so afraid the boys would fall off the mountain that I hurried to catch up with them. At one point, it was necessary to step across from one mountain to another. There was no turning back, either before or after this point. When, at long last, I reached the top, Victor and the boys were there already, and the boys greeted me with, "Mother, I never expected to see you here!" On Mt. Desert Island, only the Precipice Trail up Champlain Mountain could have been harder.

Trips by car to the top of Cadillac Mountain were made by many members of the course to be the first to see the sun rise in the USA.

Nova Scotia

Following The Bar Harbor Course, our family customarily went on to our old farmhouse in the Annapolis Valley of Nova Scotia, Canada. The Nova Scotia house played a significant role in our lives. It was a simple country house, purchased in 1842 by my great, great grandfather, Elias Bishop. My forebears were from Connecticut, coming as one of the "planter families" to hold the land for Britain following the eviction of the French Acadians. Victor and I owned this family retreat and took our children there annually. Later, geneticists frequently visited us there. Among them were John Edwards, the professor of genetics at Oxford and his wife, Felicity; Douglas Wallace and his wife, Betty; Peter Harper from Cardiff and his wife, Elaine; Rod Howell and Sally; and Judy Hall. We regularly had visits from Philip Welch and Elvira, Clarke Fraser and Marilyn, Gerry and Andree Klassen, Bill and Jean Morse, Doris and Sol Hirsch, Jim and Jean Beveridge, and various cousins who were not physicians.

We kept in touch with Victor's former fellows by Christmas cards, and they were among our closest friends.

On arrival for one of his visits to Nova Scotia, John Edwards said, "Give me a project." I handed him a bag of shards from an ancient broken dish, recently in use in the house. He very carefully glued pieces together, using epoxy cement. When he had finished and a small hole remained in the bottom of the dish, he ran out to the driveway and got a small stone to fill the hole. When the project was completed and he turned the dish over, lo and behold, written in the pottery, it said, "John Edwards, England."

Our daily routine in the Nova Scotia house included early morning tennis when it was too dark for passersby on the Palmer Road to see just how bad our tennis was. In spite of the quality of our game, it was great fun. Later, when we had improved a little, people walking the Palmer Road sometimes stopped to gawk at the two of us wielding racquets and would even cheer when we returned a shot.

After tennis, Victor wrote at a desk in the upper hall and used his computer, always working joyfully and keeping the courier packages going back to Baltimore to Della, his secretary, and to Joanna at OMIM. Victor's book, "Mendelian Inheritance in Man," had gone through 12 print editions before going on-line as OMIM.

Other Nova Scotia activities included picking blueberries or wild blackberries, the latter on our own land, or visiting the zoo our neighbors, Gail and Ron Rogerson, ran. In the evenings, we played various games with our children. Particular favorites were the dictionary game, Scrabble, Mastermind, Password, Charades, Careers, and Chronology. We tried Bridge but Victor was a somewhat frustrating partner in this since he tried to be dummy so that he could read the stack of journals on the floor beside him. We also organized beach parties at the Bay of Fundy, where the tides are the highest in the world, and cooked on a grating perched on the rocks. While the fire got started, we had snacks and someone, often Philip Welch, went the sometimes long distance over the stones, to get seawater in which to boil the corn. A salad was the first course, followed by corn on the cob. When the fire had burned low, Victor grilled steak. He then cut it on the diagonal into strips that were enfolded in slices of fresh bread. Dessert was frequently brought by the guests or was apple or blueberry pie made by me.

On one occasion in 2003, while we were in Nova Scotia, we had a telephone call from Craig Venter, who said "I will be sailing from Nantucket on Wednesday and arriving in Halifax Harbor on Saturday. Would you please charter a boat for me to go out on the Bay of Fundy and get some water samples?" His 95-ft sailing yacht was to be waiting in Halifax Harbor, while he came overland to our house, 6 miles inland from the Bay of Fundy on the other side of the peninsula. Victor and I knew nothing about chartering a boat. We could find no information in the telephone directory nor at a local tourist information booth, so we drove to Margaretsville, 21 miles away, where we found an old fisherman in his dinghy. We told him about the needed water samples, and he said he would be able to make the trip but "How far out would he want to go? and could he be sure to be there by 10 a.m.?, which would be high tide. He would have to go out and return on the same high tide," he said.

On Saturday morning, Craig arrived with two vans with scientists and photographers. We quickly drove to Margaretsville, arriving at exactly 10 a.m. Craig and a graduate student and a couple of others carrying carboys climbed down the wall ladder from the top of the pier to the small boat below in which the fisherman waited. I had some concern as they all set off to go out a mile in the small boat to fill the carboys. They did return with the water, each carboy weighing about 100 lbs.

Hoisting them to the top of the pier was accomplished by the student on top hauling them while Craig pushed from beneath. They then told us they were due back in Halifax by 3 p.m. There was no time to get to the house and give them lunch, but they all had blueberry pie in our kitchen before starting for Halifax.

Slides

On three different occasions, there were problems that concerned Victor's slides.

In 1978, we were on our way to a genetics congress in the Soviet Union. When we reached the Baltimore airport, Victor discovered that the slides, which would illustrate his lecture, were not with us and must have been left in the front hall at home. He telephoned Dora and the Richard Johns' home, which was near ours, and their son, Bob Johns, answered the phone. When told of the problem, he immediately drove to our house and picked up the slides from Dora, who met him at the front door. After a quick trip to the airport, he delivered the slides to Victor before our flight left.

On another occasion, Victor was to give an Alpha Omega Alpha lecture before dinner in Philadelphia. On the way to Philadelphia by train, Victor spread out his slides on empty seats beside him and in front of him. Then the train stopped and the conductor came through, announcing to the car that there would be a delay of one and a half hours. Victor needed to let his host know that he would be late so, seeing a gas station with a telephone booth on the other side of Route 40, he hopped off the train and crossed a field to get to it. He was just calling his host when he looked and saw the train move off. His slides and brief case were on it so he quickly called the stationmaster at Thirtieth St. Station in Philadelphia and asked him to board the train, when it arrived, and get his slides and brief case. He then caught the next train, arriving too late for the dinner but in time to give his lecture after the dinner. He was to fly, later the same evening, to Miami to lecture there the next day, which he did.

The third episode in which slides played a complicating role occurred at the Eighth International Congress of Human Genetics in Washington. Victor was the president that year, 1991, and had written out his presidential address, instead of talking from his slides as he usually did. There were 6,300 people registered for the event. As he began the talk, he called for the first slide but none appeared. He continued speaking, at times asking if there were any pictures yet. Someone near the projection booth watched as the projectionist, apparently, had a slide stuck and turned the carousel over to release it. All the slides cascaded to the floor in a large heap, unnumbered. The projectionist clasped his hand to his head and fled the projection booth. Toward the end of the address, slides began to appear at random without relation to what Victor was saying.

Special music had been composed for the occasion by Ray Sprenkle, choir director at the Second Presbyterian Church who also taught at the Peabody School of Music. The piece was entitled "Variations on a Theme" and was played by the Annapolis Brass Quintet. When they finished playing, Victor asked, "Will the composer please step forward?" No one did. Ray Sprenkle had driven from Baltimore but had been refused admission because he had no badge for the congress.

Victor used to organize trips to the top of the Johns Hopkins Dome in order to teach medical students and house staff Johns Hopkins history. At one time, Mr. Al Fayad, who owned Harrod's in London, had come to Johns Hopkins for a physical examination. Victor met him and asked him if he would like to join his trip to the top of the Dome. He accepted. It was a climb of many steps, including some up a wall ladder. After Mr. Al Fayad's completion of the climb, a planned cardiac stress test was waived. Following this, a large box used to arrive from Harrod's at Christmas time and the contents were distributed, among those working in the Moore Clinic.

Another activity that Victor and I have both enjoyed was the European School of Genetic Medicine. This was cochaired by Victor and Professor Giovanni Romeo of Bologna University. This ran for 1 week only but was modeled after the Bar Harbor Course. It has now been running for 25 years.

Among the many things important to Victor were our family and especially his "DNA-identical twin brother," Vincent, and his wife, Nancy. Also exceedingly important was our supper club, a group of nine men from Victor's time on the Osler Housestaff and their wives.

This group included:

David B. Clark and Barbara
John C. Harvey and Adele
Dudley P. Jackson and Pat
Richard J. Johns and Carol
Victor McKusick and me, Anne
Albert H. Owens, Jr. and Nan
Richard S. Ross and Boo
Gordon Walker and Betty

Two later additions were:

Richard N. Peeler and Fran
A. McGehee Harvey and "E"

They are still meeting, although in diminished numbers and somewhat sporadically. It was decided initially by the men that they would wash the dishes and for most of the years they did.

Another important membership in Victor's and my life was his membership in the American Philosophical Society, founded by Benjamin Franklin in 1743, "for promoting useful knowledge." Victor was elected to membership in 1975, and all our other engagements were scheduled around their biannual meetings.

Various honors and awards were all important to Victor including 24 honorary degrees, the Lasker Award, the National Medal of Science, and many others, culminating in The Japan Prize, the Japanese equivalent of the Nobel Prize.

Japan Prize

At the beginning of July 2007, Victor had the terribly sad news that he had a cancer of the ureter. He responded well to chemotherapy administered by Dr. Ross Donehower. Victor was able to work most of the time during his treatment. His chemotherapy was discontinued when markers in the blood indicated that it was no longer effective and to be sure that it would not interfere with the joy of his trip to Japan to receive the Japan Prize in April of 2008. This recognition was the high point of Victor's career and a marvelous experience for his family. He took a group of 12 of us, including his identical twin brother, Vincent and his wife, Nancy.

At the banquet following the presentation of The Japan Prize, Victor was seated next to her Majesty, Empress Machiko. She was most charming and asked Victor what his connection to Osler had been. Her Japanese "home" doctor in childhood in

Japan had been and was still, in 2008, a member of the American Osler Society. She also told him that she had always been interested in Nova Scotia since reading Longfellow's poem, Evangeline. She then quoted from the beginning of the poem,

> This is the forest primeval,
> The murmuring pines and the hemlocks....

She continued for more verses than Victor had memorized. She also told that in childhood, she and her 5-year-old brother had been evacuated to the north of Japan, for safety's sake. Her mother had been worried that her little brother would not be able to get enough milk to drink, so she had acquired a goat, and it had become the duty of the future empress of Japan to milk the goat. Victor had, of course, milked cows in his youth while growing up on a Parkman dairy farm.

When the banquet was over, his Majesty, Emperor Achihito went out first and waited for her Majesty, Empress Machiko, to follow. At this point, Victor, in a gesture that for him signified friendship, took her Majesty by the shoulder and said, "I want you to meet my twin brother." Our son, Victor Wayne, who had read a book on manners in Japan, was aghast. David Valle, the director of the McKusick-Nathans Institute, was amused. He had often seen Victor use the same friendly gesture in introducing a colleague. Her Majesty, Empress Machiko, was so gracious that she never indicated that etiquette had been breached.

Victor died exactly 3 months after our trip to Japan. He continued to work throughout this period, and on his last morning was focused on watching the progress of the

Bar Harbor Course, streamed to his computer in our apartment.

He planned and edited his own funeral services, one in the Second Presbyterian Church of Baltimore on August 2, 2008, and a graveside service in Parkman, ME, on August 8. At the former, there were five eulogies, each pointing out different areas in which Victor had had an impact. Our son, Kenneth, spoke for our family with "Remembering Dad." Dean Emeritus Richard Ross spoke on "Sixty-five Years at Johns Hopkins"; Margaret Budd played the organ, giving her dramatic transcription of "Finlandia" by Sibelius; David Rimoin spoke on "The Establishment of Medical Genetics"; Stephen Achuff gave "Chief of Medicine, 1973–1985"; and Clair A. Francomano spoke on "Teacher, Role Model, Mentor." Each of these speakers had been selected by Victor. The speakers selected for the graveside service were Vincent McKusick, Victor's identical twin brother, and Dr. Thomas Roderick of the Jackson Laboratory and our minister son, Victor W. McKusick, giving "Remembering My Dad."

It was common knowledge that there was some rivalry between Francis Collins of the National Human Genome Research Institute and Craig Venter of Celera, in the race to sequence the human genome first. At Victor's funeral service, each sat in the third row, on opposite sides of the church, but they both left by the center aisle, together, as if Victor had effected a reconciliation.

Victor's and mine has been a lifetime full of wonderful friendships, exciting experiences, and support for each other. I know that I speak for our children as well as for myself when I say we have been so happy to be along for the ride.

Victor standing in the dome of Hopkins

The Japan Prize 2008

Wedding photo, June 11, 1949

"With the kind permission of Frank Margeson/American Philosophical Society"

Chapter 3
The McKusick Twins: A Personal Essay[*]

Vincent L. McKusick

As Victor McKusick's identical twin (or his DNA identical brother as he sometimes called me with technical exactitude), I write here of our family background and of our bringing up together on a Maine dairy farm.[1]

In 1993, the Board of Trustees of the University of Maine System specially created The Maine Prize and conferred it upon us McKusick twins (Victor and Vincent) as its first and, to date, only recipients. At the presentation of the prize, I as the older of us twins claimed the prerogative of responding first to this shared honor. On the present occasion, however, I view it more appropriate to let my deceased brother lead off to tell of our upbringing in his own words spoken on that occasion 19 years ago. Here is Victor McKusick speaking:

> Just as I did 72 years ago, I am coming on a few minutes after Vincent. I was born about 20 minutes later and I think I was a surprise. I gather that twins were not anticipated. We were born in the morning and our father resolved that night would not fall with us unnamed. So he sat down with our sister, 13 years older, and went systematically through the section on given names in the back of the dictionary until they came to Vincent and Victor. I shudder to think of the consequences if they had had to go on to the *Z*'s; we might have ended up Zebediah and Zachariah.
>
> I am often asked if I am a geneticist because of being an identical twin – I always say "Sure." But against that theory is the fact that Vincent isn't a geneticist, and a casual survey lends no support to the notion that more twins than non-twins are geneticists.
>
> As a geneticist, I am inclined on this occasion to review what it meant for Vincent and me to be born and raised as identical twins and to think of what the genes did or did not do

[*]Note: For inclusion as an Introductory Essay in the forthcoming book, "*Victor McKusick and the Development of Medical Genetics*" edited by Dr. Krishna Dronamraju.

[1]Chief Justice, Maine Supreme Judicial Court, September 16, 1977 to February 29, 1992; Pierce Atwood LLP and its predecessor law firms, partner, 1952–September 16, 1977; of counsel, April 1, 1992–present.

V.L. McKusick (✉)
Pierce Atwood LLP, Merrill's Wharf, 254 Commercial Street, Portland, ME 04101, USA
e-mail: vmckusick@pierceatwood.com

K.R. Dronamraju and C.A. Francomano (eds.), *Victor McKusick and the History of Medical Genetics*, DOI 10.1007/978-1-4614-1677-7_3,

for us and to us. Vincent has spoken of our parents and what they gave us; that legacy included, of course, genes as well as instruction and example, and the genetic legacy was demonstrably identical.

Genetics is the science of biologic variation. When geneticists look at identical twins, it's the differences that interest them as much as the similarities. Because of essential genetic identity, any differences they see in one-egg twins are likely to be non-genetic in their causation – differences all the way from profession to physical appearance and mental temperament. The non-genetic influences that cause differences can sometimes be specified, but chance is a large element, even chance during intrauterine development. For example, Vincent and I have DNA fingerprint patterns that are absolutely identical. The likelihood of a match of these DNA patterns by chance alone is exceedingly small. On the other hand, we have different dermatoglyphic or conventional fingerprints. The variation in the skin fingerprints is because of chance differences in the development of the fingerpads in early embryonic stages. Thus, evidence taken at the scene of a crime could not distinguish us one-egg twins if the evidence was blood or semen from which DNA could be tested, but it could distinguish us if the evidence was a fingerprint on a wine glass or a door knob. Thus, the evil twin has had to be circumspect about the type of crime he committed or at least the type of evidence he left behind.

We are becoming aware of the role of chance in determining differences in other aspects of intrauterine development in addition to the fingerpads. Chance differences in the migration of neurons in the developing brain, for example, and the differences in the way they hook up may account for psychophysiologic differences, even in genetically identical twins – more about that in a moment.

Of course, chance also plays a role in the channels that extrauterine development follows. Certainly, a determining role in my going into medicine was my year-long battle with a microaerophilic streptococcus. An abscess in the left armpit and a large Meleney ulcer on the right elbow took me to the Massachusetts General Hospital for a 10-week inpatient stay in the summer of 1937, an illness from which I was rescued by the timely advent of sulfanilamide. Hitherto, I had been inclined toward the ministry; Vincent always seemed destined for the law.

We twins continued to be molded, inevitably in somewhat different directions, by our separate college and professional school experiences and by our wives who came perhaps not surprisingly, from similar backgrounds but obviously had somewhat different temperaments, personal habits, culinary preferences and favored life styles.

One rather constant non-genetic factor in our development was the 12 years of formal schooling that we got before college. We had the same experience of 8 years in a one-room schoolhouse (the so-called Pond School in Parkman) with the same teacher for 7 of those years, with only 4 students in our class and never more than a couple dozen in the whole school … One of the good things one can say for the one-room school is that able students can to some extent set their own pace. When we went on to the neighboring town of Guilford to high school, we took the same courses, the so-called college-prep course, with 4 years of Latin, 3 years of French, 3 years of mathematics but no science such as chemistry, physics or biology. English and other languages and history were emphasized. Because of the infection I missed 2/3rds of his freshman year of high school (I studied at home) but we graduated as co-valedictorians in the class of 28.

Vincent and I went to different colleges because of the prospects of doing better with scholarship support if we were not competing with each other. You can see why we are delighted that the Maine Prize is linked to a scholarship for an outstanding student. I chose Tufts in part because it had a medical school which I secretly hoped to achieve; it seemed best for pre-med training. Vincent chose Bates because of its reputation in debating which seemed a good preparation for the law.

I do not believe the complete lack of formal schooling in science before age 19 impeded me. In many ways, unjaded by earlier courses, or at least not blasé, I perhaps found the experience in college science more exciting than did some of my classmates. Genetics was

> particularly exciting, partly because of a stimulating teacher, Paul A. Warren, who incidentally came from Dover-Foxcroft, Maine and was a baccalaureate alumnus of [the University of Maine].
>
> ...
>
> In reviewing our upbringing, subtle differences in personality traits and in the way we twins behaved were discernible. It will amuse you to hear that from the very beginning and perhaps, as a result, there were differences in the ways we were treated. Our sister says, as did our mother, that Vincent was more aggressive in taking the breast, at least more vocal in his demands; I was somewhat more passive. On the farm, I was sometimes relieved of barn chores and dispatched to help our mother who at the time had no daughter at home to help her. Under the tutelage of our mother, I learned to read music, and play the old pump organ. Vincent was always more aggressively goal directed, aiming to get the job done whether it was milking the cows, getting in the hay in the summer-time or cutting down trees – a famous picture of us twins shows us at opposite ends of a two-man crosscut saw; Vincent is clearly pushing as well as pulling. Vincent was always more argumentative than I; I prided myself on being patient and resigned in the face of a prolonged illness.
>
> Our pre-college education was adequate mainly because of the home situation. Today, we expect the schools to do it all. The isolation in which we twins grew up sheltered us from some iniquitous influences of urban living. We worked closely with our parents who put high value on both industry and learning. We were voracious readers. To this day, I am almost a pathologically compulsive reader. Many times our father would say to one of us, "Bring your book (it was always assumed we had one) and watch the cows", referring to the times that the cows were let out into the fields to graze on the second growth after haying. We always had projects. The school year had 32 weeks, running from late August to mid June, with a 10 week vacation in mid winter when the snow was deepest and a 10 week vacation in the summer. The 10 week winter vacation was used for projects which was, for example, a detailed family genealogy for me and charting of the Kings and Queens of England for Vincent, both perhaps presaging future careers.

All of the above remarks made by Victor 19 years ago I find squarely in line with my memory of today. Consequently, I here make my role one of emphasis and supplementation only.

Victor and I incurred many debts to others for getting us to success in life. Foremost is our debt to our parents, Carroll and Ethel Buzzell McKusick. Our greatest debt to them may well have been genetic, but Victor of course could better assess that than I. However, I *do* know the magnitude of our nongenetic debt to them: for their raising us, in those prewar depression years, on a family dairy farm in the town of Parkman, deep in rural Maine, and especially for their being the kind of people they were.

Both came from the families that first settled the Parkman area in the first half of the nineteenth century. All four sets of Carroll's great-grandparents lived within a single 10-mile area. Carroll's father kept a small farm, and during winters, cooked in lumber camps; Ethel's father ran the general store and post office. Carroll was the first in his whole line to go to college – graduating ØBK from Bates College in 1904 (he always called it "Aught Four"). A severe bout with rheumatic fever at age 18 frustrated Ethel in her ambition to study dramatics at Emerson College in Boston, but that did not prevent her, right out of high school, from teaching in the local one-room grammar schools. From those educational backgrounds, Papa, as we called him, was ever the Latin, Greek, and math scholar; Mama ever the writer, the drill-mistress in diction and basics, and the public speaking coach. They were natural teachers.

In 1906, Carroll took his new bride to Chester, Vermont, where he was the high school principal, but in 1913, when they had two children under the age of 5 and a third soon to come, Carroll's poor health sent them back to their roots in Parkman, Maine. By the time we twins came along 8 years later, they had started dairy farming and were steadily building what became a prize herd of purebred Jerseys.

Until we twins went to college in 1940 – separately to Bates and Tufts – family completely dominated our universe, economically, socially, and in every other way. The farm was a joint family enterprise with hard work expected from all as a matter of course – particularly from two healthy teenage boys who did not go to college until almost 19 – work that involved care of the dairy cows and milking them by hand, haying, delivering milk and cream on a retail route in Guilford, even cutting wood and peeling pulp. Until our older brother Robert, a high honors graduate of the University of Maine, returned in the late 30s to join our folks in operating the farm, it had neither a tractor nor a milking machine. Equine and human labor took their place. As you can imagine, that hard farm work added considerable attractiveness to higher education as the road to greater opportunity. The 30s were tough years for many, but our Maine farm sheltered us from the depression. As Dwight Eisenhower said of his Kansan youth, "We were poor, but we didn't know it."

Our outside social life was with the family – in the Baptist church, the grange, and the schools and their activities. Nothing more. I remember going to the movies only once before college. Our entertainment was at home – a radio, newspapers and magazines, and many books, especially Maine books. Nights after the chores were done, Papa would, for example, read aloud the Kenneth Roberts novels as they appeared starting with "Rabble in Arms." And over all the years we supplemented the local library by going through the two bookcases of high school texts that years before book salesmen had left as samples with the high school principal.

Our parents also defined our aspirational universe. They instilled in us great family pride – pride of our forebears (Mama was the family genealogist), pride in the accomplishments of our much older siblings – Grace, 13 years older than the twins, was the first woman from Parkman to go to college, graduating from Bates ØBK to become a much-respected high school teacher and minister's wife in Maine; Harry, 10 years older, having proved himself as a foreman in the CCC, became a career forester for the State of Connecticut; and Robert, 7 years older, took top honors at the University of Maine before returning to take over the family farm. Scholastically, our family expected us twins of course to match the high school and college records of our parents and older siblings. Relevant to Victor's future genetics career, we twins participated in keeping records of the milk production of our herd of purebred Jersey cows. Also relevant to his future career, Victor while still in high school joined his farmer brother Robert in authoring a paper explaining the fawn color of most Jersey cattle as a recessive trait. They sent it off to *The Journal of Heredity*, but never received even an acknowledgment. Happily, that experience did not stifle Victor's drive to write and publish.

Spending our first 19 years so closely with our parents, we twins inevitably picked up some of their personality traits – if we had not already inherited them. We harbored a hope (which Victor is said to have realized better than I) that we acquired some of Mama's remarkable equanimity. Her role as farm wife and mother was not an easy one. The farmhouse, although historically distinguished as the oldest in town, dating from 1816, was barren of modern conveniences of central heat and modern bathroom facilities. Electricity did not arrive until we twins were 8 years old. A large, handsome woman, Mama suffered from a rheumatic heart and bad varicose veins, conditions not helped by having brought two oversized babies into the world when she had already passed Jack Benny's unchanging age of 39. But despite the demands of farm life and despite her poor health, she was at all times even-tempered and of good cheer. She was always optimistic; hard work and perseverance would resolve any problem and achieve any goal. She died in 1944 at age 62 – shortly after Victor and I were out of college.

Both of our parents were deeply interested in public affairs and assumed we would be also. More than once, Mama proudly told us of being the very first woman in the town of Parkman to cast her ballot in the first State election that came after the adoption of the 19th Amendment in 1920. Papa held the whole range of town offices and was somewhat active in politics – Republican of course – but not until Robert returned from the University to take over the farm was Papa able to go to Augusta for 12 years of service in the Maine House and Senate and then for 17 years (until his death at age 87) of membership on the State Board of Education with two terms as its chairman.

I end on a very personal note. Contrary to the conflict sometimes reported between twins, Victor and I each for nearly 87 years counted the other as his number one best friend, cemented by genetic fact. Typical of our closeness and of Victor's generosity, he insisted less than 3 months before his death that my wife Nancy and I accompany him and Anne to Tokyo for the presentation to him of the Japan Prize. One can truly say that genetics meant a very great deal to Victor.

The twins at about 18 months

The twins at perhaps 10 years

The twins at about 5 in front of the farm barn

The twins at about 9 with two neighbor friends on one of our farm workhorses

The twins at about 6 with their sister Grace

Both couples at the farm in Cape Elizabeth, Maine

Twins at the farm

Twins at Carroll L. McKusick School

Chapter 4
Personal Recollections of Victor

Krishna R. Dronamraju

It is with mixed feelings of sadness and pleasure that I record some recollections of Victor. He was a man of exceptional loyalty and kindness toward his close associates. This was not immediately obvious as he was a man of few words and not given to much small talk. However, I came to appreciate fully his admirable qualities as I got to know him over 40 years of friendship.

When my wife and I first arrived in Baltimore, Victor took special care to make us feel comfortable. At first, he put us up in the rooms under the Hopkins Dome. Only later, it dawned upon me that it was an honor accorded to very few. Afterward, he drove us around Baltimore for several hours to look for an apartment, while his colleague Tony Murphy followed us in his car. It was a little overwhelming to receive so much personal attention.

Although our relationship was quite formal at Hopkins, I enjoyed Victor's warmer and genial side elsewhere, when we traveled to meetings or during the Bar Harbor courses and also during the Baltimore Harbor cruises including crab cakes and beer after March of Dimes conferences which were carefully arranged by Victor. I recall one occasion when we traveled together to a meeting at Columbia University in New York. Everything went well and businesslike, but there was always time for a glass of beer and polite conversation at the airport before returning home. There was also the occasion, I was told, when Robin Bannerman and others phoned Victor from the next room, while watching his reaction, saying that the caller was J.B.S. Haldane who was desperately seeking Victor's help because the American authorities would not let him leave Baltimore airport! Apparently, Victor rose from his chair with great concern and consternation. It took him a while to appreciate that joke and regain his composure and humor.

K.R. Dronamraju (✉)
Foundation for Genetic Research, P.O. Box 27701-0, Houston, TX 77227, USA
e-mail: kdronamraj@aol.com

K.R. Dronamraju and C.A. Francomano (eds.), *Victor McKusick and the History of Medical Genetics*, DOI 10.1007/978-1-4614-1677-7_4,

I came to realize rather slowly that he was a great admirer of my mentor, J.B.S. Haldane, when one day he told me that he used to read Haldane's storybook, *My Friend Mr. Leakey*, to his own children. He sent me the following note, dated December 21, 2006.

> Dear Krishna,
>
> I have been fascinated to read J.B.S. Haldane's "My Friend Mr. Leakey", first published in 1937. Did he publish other children's books or fiction? Am I correct that he had no children? Did he enjoy playing with nephews or nieces or other children?
>
> All the best,
> Victor

I replied "no" to the first question and "yes" to the other two. I told him that Haldane was, in fact, very popular with children.

Then there was the amusing occasion when one of the visiting Liverpudlian doctors from England came to say good-bye to Victor at the end of his 3-year stint. Victor got up from his chair, shook hands, said "good-bye," and instantly returned to his work without saying another word, leaving the poor fellow standing there speechless for several minutes!

When I proposed him for the National Medal of Science, he provided me with much biographical information. He was awarded the medal by President George Bush in a White House ceremony in 2002 and was extremely proud of that honor. He had the Christmas cards specially made that year with a photo of him receiving the medal from the President. I was amused and proud when he sent me one of those cards with a note: "Pardon my immodesty…."

With his interest in multiple disciplines and their interaction, it is not surprising that Victor was especially interested in two of my books: *Haldane's Daedalus Revisited* (ed), (Oxford University Press 1995) and *Infectious Disease and Evolution* (ed) (Cambridge University Press 2004), both emphasizing history, genetics, evolution, and ethics.

Serendipity brought us together on one occasion in Oxford, England. As my wife and I were leafing through the books at Blackwell's, our pride and joy in Oxford, suddenly, we heard my name shouted across the room. We spun around with alacrity. Victor and Anne just entered the premises and were obviously very happy to see us. We had the advantage; we knew the McKusicks were in town, while they didn't know we were. We continued our conversation over an excellent lunch at Randolph's. Anne told us, "Isn't it incredible, just 2 days ago Victor and I were picking blueberries on a hill in Nova Scotia and here we are with you!"

In Baltimore, we used to dine occasionally at the Johns Hopkins University's Faculty Club with Victor and Anne. The last time we saw them was in 2007 at lunch at an Indian restaurant with Aravinda Chakravarti and his wife Shukti, both colleagues at Hopkins. It was a most enjoyable occasion for all of us. Little did we know that it would be the last.

I remember vividly his enthusiasm and hard work, his readiness to listen to new ideas, his concern for ethical standards in research, his wide scholarship, and especially his decent and kind attitude toward colleagues and friends.

J.B.S. Haldane and Mrs. Haldane (Helen Spurway)

Lunch with Victor and Anne McKusick, (standing behind) Aravinda Chakravarti, Shukti, Krishna Dronamraju, Michele Wambaugh

Krishna and Michele congratulating Victor for the National Medal of Science Award, 2002

Victor in Houston with Krishna Dronamraju and Frank Arnett

Chapter 5
Memories of Victor McKusick

Don M. Long

Victor McKusick and I were appointed as departmental directors at Johns Hopkins 1 year apart. Thus, we were presented to the Biennial Alumni Meeting together as the newest chairs at Johns Hopkins. As would be typical for neurosurgeons, I was delayed in the operating room and reached the biennial banquet too late to be seated at the table in the front without disrupting the proceedings. I found myself with a group of internists celebrating a class reunion. They really did not know who I was, so when Victor was announced as the new chair of the great Department of Medicine at Johns Hopkins, they were free with their comments. All commented on what a fine choice he was. There was a general approval of his choice as the new chair, but it was colored by a reservation. One member of the group summarized the general feeling by saying Victor McKusick is a wonderful person and a wonderful choice. It is just too bad that his clinical interests are genetics and not something important like cardiology. It would be interesting to know what comments they might have today after Victor's accomplishments are recognized worldwide, and his seminal contributions to medical genetics, the greatest evolution in medicine of the past half-century, are world famous.

My personal experience with Victor as a clinician began in the early 1980s. Victor called me one day to ask what I knew about neurological symptoms in patients with Marfan's syndrome. I had to truthfully answer nothing. He then told me about a small group of patients he had seen who had back pain and radicular complaints in association with large dural ectasias. He wanted to know if I could find any other cause of their back and leg pain and often associated neurological deficit.

He asked if I would be willing to see a few of these patients for him. He referred two or three patients for evaluation, and we concluded together that the dural ectasias must be the cause of both the back pain and the radiculopathies in these

D.M. Long (✉)
Department of Neurosurgery, Johns Hopkins Hospital,
Baltimore, MD 21204-3741, USA
e-mail: dmlong@jhmi.edu

K.R. Dronamraju and C.A. Francomano (eds.), *Victor McKusick and the History of Medical Genetics*, DOI 10.1007/978-1-4614-1677-7_5,

patients because I could find no other reasonable cause. In typical fashion, Victor encouraged others of his group to pursue this clinical problem, and they have published the majority of the papers on this entity.

My own involvement continued because several of these patients had either initially severe or subsequently progressive neurological deficits which were important. Several had intractable pain. The typical patient was complaining of back pain, had a well-defined lumbar or sacral radiculopathy, and often had an associated neurological deficit. The usual patient had Marfan's syndrome, but a few were diagnosed with Ehlers-Danlos syndrome. The typical ectasias were around lumbar and sacral roots, usually L5 and below. They usually traversed the sacrum, often thinning it dramatically. Many presented as abdominal cysts. The largest of these cysts in our small series held 2.5 l of fluid. Intractable pain, neurological deficits, and progression of neurological deficits occurred in this small group of patients. We undertook surgical repair in those with serious or progressive deficits. The surgery proved to be a formidable undertaking, and while we were successful in all, more than one surgery usually was required to obliterate the large connections between spinal subarachnoid space and nerve sheath. Most patients were treated with posterior obliteration of the orifice of the ectatic defect, and some required additional transabdominal procedures, usually by endoscopy, to open a residual cyst.

There still are no epidemiological studies within the connective disorder syndromes to allow us to determine the frequency with which these symptomatic ectasias occur. The surgical techniques developed so far are not fully satisfactory. Nevertheless, it has surprised me that so few people seem to have ever heard of this observation of Victor's and several of his collaborators. I still see patients from major centers, where no one has truly correlated the dural ectasias with the patient's complaints or offered treatment.

This apparently small but important group of patients helps illustrate a point I want to make about Victor McKusick. He was a superb clinician and meticulous in his correlation of symptoms and signs with possible causes in his patients with genetic diseases.

His correlation of the large ectasias with serious symptoms is not widely appreciated. It is another example of an important correlation which Victor made between the complaints of patients and their genetic issues.

At a time when genetic medicine is having difficulty with translation of the great advances in genetics to clinical practice, this is a simple example of how such a translation can occur effectively.

Chapter 6
Memories of the Moore Clinic, 1960–1965

David J. Weatherall

Introduction

Medical genetics was born at the beginning of the twentieth century following the work of the English physician Archibald Garrod [1]. His studies of biochemical individuality were stimulated by his observations on patients with alkaptonuria, and in his early description of the disease, he emphasized the frequency of consanguinity in the parents of his young subjects. William Bateson, one of the early supporters of Mendel's work, pointed out to Garrod that this was exactly what would be expected if these diseases followed a recessive form of inheritance. Garrod collected more inborn errors of this type and in June 1908 delivered the Croonian Lectures to the Royal College of Physicians entitled '*Inborn Errors of Metabolism*'. The lectures were published under the same title as a monograph in 1909 [2]. In 1931, Garrod expanded these concepts in his second, and even more remarkable, book *Inborn Factors in Disease* [3]. In this neglected work, he expands the concept of inherited biochemical individuality combined with interaction with the environment as the basis for disease.

Garrod's remarkable discoveries had virtually no immediate impact on the medical profession. Indeed it is clear from the reviews of the books that there was very little understanding of his message at the time. Although some other disorders with a Mendelian pattern of inheritance were discovered in subsequent years, medical genetics as a branch of medicine did not move forward at all until after World War II.

D.J. Weatherall (✉)
Weatherall Institute of Molecular Medicine, John Radcliffe Hospital,
University of Oxford, Headington, Oxford OX3 9DS, UK
e-mail: liz.rose@imm.ox.ac.uk

K.R. Dronamraju and C.A. Francomano (eds.), *Victor McKusick and the History of Medical Genetics*, DOI 10.1007/978-1-4614-1677-7_6,

There is no doubt that Victor McKusick was one of the major founding fathers of the new specialty of medical genetics, a field which started to evolve in the mid-1950s. Interestingly, McKusick and Garrod have a great deal in common with respect to how they developed their careers which, in both cases, remained firmly embedded in general clinical medicine. Garrod remained a practising general physician throughout his career and his final post was Regius Professor of Medicine at the University of Oxford. McKusick, after completing his house staff training at Johns Hopkins Hospital, Baltimore, in 1952, trained in general medicine and cardiology and for some years developed the field of spectral phonocardiography. Like Garrod, McKusick took unusual observations from the bedside to the research laboratory. His cardiac practice included the study of groups of patients with inherited disorders of connective tissue, and within 4 years, he had written a classic textbook on this topic, *Heritable Disorders of Connective Tissue* [4]. Although largely a self-taught geneticist, McKusick was greatly influenced by other geneticists Bently Glass at Johns Hopkins University and Curt Stern at the University of California (Berkeley). It was gradually becoming clear that medical genetics was here to stay, and in 1957, he was invited to develop a Division of Medical Genetics, based in the Chronic Disease Clinic which had been established earlier by J. Earle Moore. The name of the Moore Clinic was to become synonymous throughout the world with medical genetics over future years.

At the present time, when medicine is characterized by intense sub-specialization, and medical genetics has become separated from the mainstream of other clinical specialties, it is interesting to reflect on the critical importance of the way in which the careers of Garrod and McKusick remained embedded in the mainstream of general medicine. In reviewing McKusick's research activities over many years, one is struck by their enormous breadth.

Working in a large hospital with many excellent specialist departments, he was able to collect and analyse genetic conditions stemming from almost every sub-specialty of medicine and most clinical disciplines. Indeed, he often described his major field of interest as 'genetic nosology', that is, the definition of discrete genetic entities. And like Garrod, later in his career he occupied a prestigious Chair in Medicine; in 1973, he was appointed Chairman of the Department of Medicine and Physician-in-Chief at Johns Hopkins Hospital, a post that he held until 1985.

In 1990, McKusick was awarded the Kober Medal, the highest honour of the Association of American Physicians. He invited me to give the presentation address, which I finished with the following statement: 'As well as the development of the tools of quantitative and molecular genetics, progress in human genetics required a truly renaissance figure with the breadth of vision and ability to describe and catalogue the enormous variety of phenotypes that constitute the diverse body of genetic diseases of Man. Without doubt Victor McKusick has been that person; the genetic nosologist supreme' [5].

In this short appreciation, I will first offer a few reminiscences of my period working with McKusick in the Moore Clinic and then outline how these early studies developed later towards the beginnings of an understanding of the molecular basis for the thalassaemias and related disorders, the first Mendelian diseases to be defined at the molecular level.

The Moore Clinic in 1960

At the end of 1957, after qualifying in medicine in Liverpool and 2 years postgraduate experience, I had to serve 2 years of national service in the British army. In its wisdom, it sent me to work in Southeast Asia, for 1 year in Singapore and a second year in Malaya (as it was then known). While working in the children's ward at the British Military Hospital in Singapore, I came across a Nepalese baby with profound anaemia who was being kept alive by blood transfusion in the absence of any diagnosis. Her family had come down from Nepal with her father who was serving in a Gurkha regiment. After a lot of work and with the help of a biochemist, Frank Vella from Singapore University, we established that this child had thalassaemia, a rather unexpected finding since at that time this disease was thought to be restricted to the Mediterranean region.

Although very little was known about thalassaemia at that time and the structure of haemoglobin had just been determined, there were hints that thalassaemia, like sickle cell anaemia, might be a genetic disorder of haemoglobin production. I therefore wrote to my mentor in Liverpool, Cyril Clarke, and said when I came out of the army that I wanted to go somewhere where I could be trained in genetics, haematology and protein chemistry. He thought that if I came back to England with this requirement, I would be sent for psychiatric assessment!

Because of his own evolving genetic interests, he had built up contacts with McKusick, and he thought that I had best go as a research fellow to the Moore Clinic. He had already sent one fellow from Liverpool to the Clinic, and since his fellowship was coming to an end, McKusick was able to give me a place and I arrived in Baltimore in 1960.

Although the clinic had only been established for a few years, it was already humming with activity and its staff and fellows were carrying out research in widely diverse aspects of human genetics under the critical eye of McKusick. It was suggested that I join Ned Boyer, who had worked as a cardiology fellow with Victor in 1956 and, after a year's sabbatical at the Galton Laboratory in London, had switched to biochemical genetics. Tony Murphy, a biostatistician who had come to Baltimore from the Hammersmith Hospital in London, was applying his skills to population genetics and was extremely helpful to each new fellow as they arrived, helping McKusick to settle them in to the Baltimore environment. Malcolm Ferguson Smith and his colleagues were developing a very strong cytogenetics laboratory, and my predecessor from Liverpool, David Price Evans, had already established some elegant work in the field of pharmacogenetics. Jim Renwick was an early pioneer in the development of computer programmes to study genetic linkage. Apart from these British fellows, there were at least 12 other fellows from different countries working in the clinic in diverse areas of medical genetics.

Most of the fellows helped in the various genetic clinics attached to the Moore Clinic, and we met regularly to present our work to McKusick. One of the most memorable experiences of that time was the monthly journal club which was held at McKusick's home. Each of us was given specific journals to cover, and we were expected to complete index cards with summaries of papers of genuine genetic interest.

McKusick used these cards to help him develop annual reviews of genetics which were published in the *Journal of Chronic Diseases*. This was the beginning of what led to one of Victor's major contributions, *Mendelian Inheritance in Man* [6].

This work, in which he arranged lists of genetic disorders under the headings of autosomal dominant, autosomal recessive and X-linked, was first published in 1966 and then regularly updated and edited; after 12 editions, it moved to an online version which has continued to this day.

Looking back over this period in the Moore Clinic, it is remarkable to reflect on how many of those who were trained by Victor later became international leaders in different fields of medical genetics all over the world (Table 6.1). I have tried to summarize Victor McKusick's extraordinary personal contributions to different

Table 6.1 Later careers of some of Dr. McKusick's fellows

Name	Later career, Title	School, Location
Beighton, Peter	Professor of Human Genetics	U. Cape Town, S. Africa
Bias, Wilma B.	Director of Immunogenetics Professor of Medicine	Johns Hopkins
Borgaonkar, Digambar G.	Director of Cytogenetics	Medical Center of Delaware
[a]Bowen, Peter	Professor of Pediatrics	U. Edmonton, Alberta
Boyer, Samuel H., IV	Professor of Medicine	Johns Hopkins
Brunt, Peter W.	Consulting physician and physician to the Queen in Scotland	Aberdeen, Scotland
Bundey, Sarah E.	Consulting medical geneticist and neurologist	Birmingham, England
Char, Florence	Professor of Pediatrics	U. Arkansas
Conner, J. Michael	Director, Duncan Guthrie Institute of Medical Genetics	Glasgow, Scotland
Danks, David M.	Director, Murdoch Institute	Melbourne, Australia
Donahue, Roger P.	Cytogeneticist	U. Miami
Emery, Alan E. H.	Chairman, Dept. of Human Genetics	Edinburgh U., Scotland
Ferguson-Smith, Malcolm A.	Chairman, Dept. of Pathology	Cambridge U., England
Filippi, Giorgio	Professor of Medical Genetics	Trieste U., Italy
Kelly, Thaddeus E.	Professor of Pediatrics	U. Virginia
Kunkel, Louis M.	Chief, Div. of Genetics	Children's Med. Ctr., Boston
Glynn, Michael	Haematologist	Toronto
Goldberg, Morton F.	Director, Wilmer Ophthalmological Institute	Johns Hopkins
[a]Goodman, Richard M.	Professor of Medical Genetics	Tel Aviv U., Israel
Hall, Judith G.	Professor of Medical Genetics	U. British Columbia, Vancouver
	Chairman, Dept. Pediatrics	
Harper, Peter S.	Chairman, Dept. Medical Genetics	U. Cardiff, Wales
Jorgensen, Ronald J.	Professor of Dental Genetics	U. Texas, San Antonio
Laberge, Claude	Professor of Medical Genetics	Laval U., Quebec
Lee, Catherine	Cytogeneticist	Edmonton, Alberta

(continued)

Table 6.1 (continued)

Name	Later career, Title	School, Location
Leeming, James M.	Consulting geriatrician	England
[a]Levin, L. Stefan	Associate Professor	Johns Hopkins
Lie, Sverre O.	Professor of Pediatrics	Oslo, Norway
Maumenee, Irene Hussels	Professor of Ophthalmology	Johns Hopkins
Murphy, Edmond A.	Professor of Medicine	Johns Hopkins
Mutalik, Gururaj	Professor of Medicine and Dean	Poona, India
Niermeijer, Martinus F.	Professor of Medical Genetics	Rotterdam
[a]Pascasio, Flora	Geneticist	Manila, Philippines
Patterson, Donald F.	Professor of Veterinary Medicine	U. Pennsylvania
Penchaszadeh, Victor B.	Chief, Medical Genetics	Beth Israel Hosp., N.Y.
Penttinen, Risto P.	Professor of Medical Chemistry	Turku, Finland
Pierce, Edward R.	Dean	U. Delaware
Pope, F. Michael	Medical Research Council	Mill Hill, U.K.
Porter, Ian H.	Chairman, Dept. of Pediatrics	Albany Med. Sch., N.Y.
Premchai Yuniband	Medical Geneticist	Bangkok, Thailand
Price-Evans, David A.	Chairman, Dept. of Medicine	Liverpool, England
Rimoin, David L.	Professor of Pediatrics Chairman, Dept. of Pediatrics Cedars-Sinai Med. Center, L.A. CA	UCLA Sch. of Medicine L.A., CA
Sayli, Betki S.	Professor of Human Genetics	Ankara, Turkey
Schimke, R. Neill	Professor of Medicine	U. Kansas
Scott, Charles I.	Professor of Pediatrics, Jefferson Medical School, Chief of Genetics, duPont Institute	Wilmington, Del.
Sherwin, Roger W.	Professor of Preventive and Social Medicine	U. Maryland
Siggers, David C.	Director of Medical Genetics	Southampton U., U.K.
Skovby, Flemming	Associate Professor of Pediatrics	Copenhagen, Denmark
Srsen, Stefan	Professor of Medical Genetics	Martin, Czechoslovakia
Stoll, Claude	Professor of Medical Genetics	Strasbourg, France
Suh Soon Kyu	Professor of Medicine and Dean, Seoul U. School of Medicine	Seoul, Korea
Temtamy, Samia A.	Chief of Medical Genetics	Cairo, Egypt
Walker, Brian	Consulting physician	Liverpool, England
Weatherall, David (Sir)	Nuffield Professor of Medicine	Oxford U., England
Welch, J. Philip	Professor of Medical Genetics	Dalhousie U., Halifax, Nova Scotia
Wells, R. S. ("Charles")	Dermatologist and geneticist	Guy's Hospital and St. John's Hospital, London, England
Wise, David	Consulting physician	London, England
Woodrow, J. C.	Professor of Medicine	U. Liverpool, England
Wuu, Kuang-dong	Director, Institute of Molecular Biology	Taiwan
Yoder, Omer C.	Representative of National Cancer Institute	Brussels, Belgium

This table was compiled by Victor McKusick as part of the presentation ceremony for his award of the Korber Medal [5]

[a]Deceased

Fig. 6.1 The McKusick family off the coast of Maine, August 1967. Victor, Anne, Carol-Ann (aged 14), Kenneth (aged 4) and Victor Wayne (aged 3). This photograph was provided to the author by Victor McKusick for his presentation address for the Korber Medal to McKusick in 1990

areas of medical genetics elsewhere [7], and he has written several historical accounts of his work [8, 9]. This period of development of the field in the Moore Clinic undoubtedly had a major international effect on the evolution of medical genetics from the 1960s onwards, reflecting the activities of those who were trained there.

What kind of a man could pack this extraordinary activity in the development of medical genetics into one lifetime and yet remain so close to medical practice that later in his career he became chairman of Medicine at Johns Hopkins University? One factor undoubtedly was the support of his wife Anne, a rheumatologist at Hopkins, and their three children (Fig. 6.1). It was rumoured that he went to bed quite early but got up at about 5.00 am to start writing papers. One of my earliest memories of McKusick seems to epitomize this approach to his work. Some of my early work on haemoglobin genetics went rather well a few months after I had arrived at the Moore Clinic in 1960 and, completely unbeknown to me, McKusick and Boyer offered it for a presentation at the 74th Annual Meeting of the Association of American Physicians in Atlantic City in May 1961. Remarkably, this paper was placed early on the plenary session on the first morning of the meeting [10]; I had never heard of the Association and nobody had had the foresight to warn me about what was in store when I presented my first scientific paper in public.

As I sat in my hotel room the night before the event, the telephone rang. 'Victor here; would it be all right to meet for breakfast at 6.30 am tomorrow?' We met at 6.30 am, completed our breakfast within half an hour, and then walked along the waterfront and entered the Steel Pier Theater, a vast building that was completely empty. McKusick marched briskly to the front row of the theatre, sat down, opened his case and started writing. After an hour or so, I plucked up courage and asked him what we were doing there. 'This is where you are speaking' said Victor, and continued to work on his paper. Some hours later, as I rose to face the combined might of the Association, trembling with terror, I glanced down at McKusick. He was still writing at a furious pace; the second paper of the morning was going well.

I doubt if McKusick had much time for many interests outside medicine. A few years after I had returned to the UK, I received a beautiful Christmas card from him, a card of such artistic elegance that I wondered if, at last, he had started to take time off to sample the better things of life. I ought to have known better; a closer inspection of the lovely Madonna pictured on the card revealed that her child had an extra digit on both hands. The Madonna had been based on a picture of an Amish mother with a child affected by the Ellis-van Creveld syndrome! Another British fellow told me that he had been invited to the McKusick home after arriving in Baltimore and had asked Victor if he could sample some American television. He apologized saying that although they had bought the set 2 years ago, he still did not know how to switch it on.

Early Studies of Haemoglobin Genetics and Thalassaemia in the Moore Clinic

In the 1950s, Oliver Smithies had developed a more sensitive approach for separating and analysing proteins by electrophoresis on starch gel. Ned Boyer had learnt this technique while working with Harry Harris in London and was using it to look for variant forms of alkaline phosphatase when I joined him in 1960. With the help of a delightful Estonian technician, Linda, he was running a factory with this technique, carrying out analyses on over 20 gel trays a day! Because I wanted to work on haemoglobin genetics, we wondered if this technique would be valuable in this respect. At the time, it was known that human adult haemoglobin consists of two α chains and two β chains (α2β2), foetal haemoglobin has two α chains and two γ chains (α2γ2), and that the minor component of human adult haemoglobin, haemoglobin A2, has two α chains and two δ chains (α2δ2). What was not known was whether the same α globin genes were involved in the production of each of these haemoglobins or whether there were different sets of α genes involved in directing α globin synthesis in foetal and adult life. We hypothesized that if the same set of α genes were involved at all stages of development, then those who had inherited a structural haemoglobin variant due to an abnormal α chain would produce both abnormal foetal and adult haemoglobins, whereas if there were separate sets of α genes for foetal and adult haemoglobin, those who had inherited an abnormal α chain gene would either produce abnormal foetal or abnormal adult haemoglobin but not both.

My first task therefore was to collect a large number of samples from the Baltimore Afro-American population to screen for haemoglobin variants. We then evolved a system for defining whether they were α- or β-chain variants by dissociation/reassociation experiments analysed by starch-gel electrophoresis (Fig. 6.2). The outcome of this work showed clearly that α globin synthesis occurs under the direction

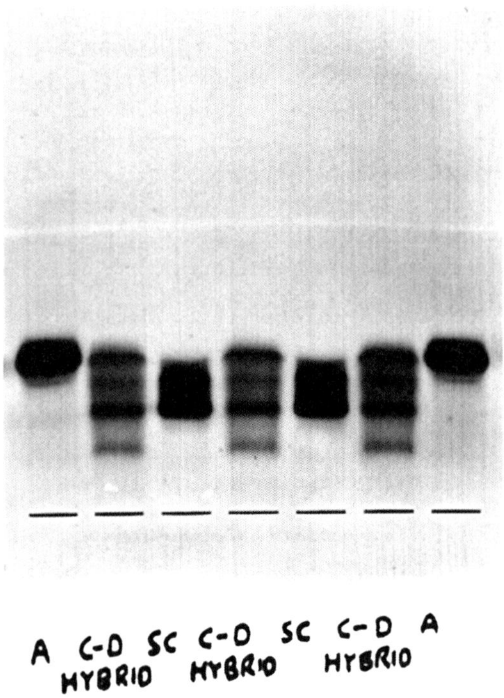

Fig. 6.2 Early dissociation/reassociation experiments carried out in the Moore Clinic in 1960. These experiments were carried out to determine whether the common form of haemoglobin D in the Afro-American population of Baltimore was an α or β chain variant. Mixtures of either haemoglobins S and C, which are both β chain variants, were compared with those of haemoglobins D and C. The mixtures were subjected to a low pH at which the globin chains dissociate, and then the chains were allowed to reassociate at an alkaline pH. In the case of haemoglobins S and C, only the original variants appeared on electrophoresis. However, because the haemoglobin D was an α chain variant, when mixed with haemoglobin C, after dissociation and reassociation, four haemoglobin variants were found: α2 Aβ2 A, α2 Dβ2 A, α2 Aβ2 C and a hybrid molecule α2 Dβ2 C. These variants were separated by starch-gel electrophoresis, pH 8.4. The samples at either end of the gel are haemoglobin A markers

of the same set of genes in both foetal and adult life, indicating that disorders of α globin production would be expressed throughout development, unlike those affecting β chain production which would only occur after birth. Similarly, it became possible to distinguish between heterozygotes for α or β chain variants because the former carried two forms of haemoglobin A2, reflecting normal α chains combined with δ chains and the abnormal α chains combined with these chains [10, 11]. The structures of the abnormal foetal haemoglobins found in babies who inherited α haemoglobin variants were later confirmed by more extensive structural analysis in studies in collaboration with Corrado Baglioni [12].

While analysing the pattern of haemoglobin in the cord blood of large numbers of Afro-American babies in Baltimore, I was surprised to find that up to 5% of them carried a rapidly migrating haemoglobin variant which later structural studies showed was haemoglobin Bart's. A similar variant had been observed in studies in West Africa, but it was not clear whether this observation was of any clinical significance. Haemoglobin

Bart's has the structure γ4 and is the foetal counterpart of haemoglobin H, which has the structure β4. It was thought that these variants might reflect defective α chain production due to α thalassaemia, but the strange thing about the Baltimore babies was that those who had had Hb Bart's at birth did not have HbH as they developed.

However, they did have thalassaemic red cell changes. Furthermore, extensive family studies revealed similar changes in some of their relatives although no haemoglobin H on electrophoresis. These observations suggested that there was a mild form of α thalassaemia which is particularly common in African populations and that perhaps it can be most easily recognized in the neonatal period at a time when both β and γ chains are competing for α chains and hence when a mild deficiency of α chains might become more obvious [13]. This hypothesis turned out to be correct when, many years later, it was possible to analyse the α globin genes at the DNA level and show that there is a common form of α thalassaemia in African populations which is due to a deletion of one of the pairs of α globin genes.

Later Work on Thalassaemia and Human Haemoglobin in Baltimore

Because I needed training in haematology and protein chemistry, I spent my later years in Baltimore working in the Department of Hematology under C. Lockard Conley and the Department of Biophysics under Howard Dintzis, although keeping in regular touch with the Moore Clinic.

By the early 1960s, it was becoming clear that further progress towards an appreciation of the molecular basis of the thalassaemias would require a better understanding of the pattern of disordered haemoglobin synthesis in these conditions. At this time, the patterns of haemoglobin synthesis were being investigated in animal models in which the animals had been treated with phenylhydrazine to induce a significant hemolytic anaemia and high reticulocyte count such that it was possible to follow the patterns of synthesis using the incorporation of radioactive amino acids into the globin chains.

However, even at that time, when ethics committees were not so active as they are today, it was not possible to treat human beings with phenylhydrazine and therefore I had to struggle away to find the conditions in which to do similar studies in humans with much lower reticulocyte counts. In the end, I was able to obtain linear synthesis of haemoglobin in the test tube for up to an hour and the next problem was how to separate the globin chains.

At that time, there were two protein chemists from Cambridge working in the Department of Biophysics, Michael Naughton and John Clegg; the latter was to become my lifelong research partner and colleague. With John's help, we developed a way of separating the globin chains and were able to show quite unequivocally that thalassaemia is not so much a disease of defective haemoglobin synthesis as a disorder of imbalanced globin-chain synthesis, with excess α chains produced in β thalassaemia and excess β chains in α thalassaemia [14]. This method was used widely throughout the world later and was successfully applied to the prenatal diagnosis of the thalassaemias in the 1970s.

Later, this approach was further modified so that by short periods of pulse labelling of the α or β globin chains it was possible to measure the rates of initiation of their synthesis, their patterns of elongation and the rates of termination. In those forms of thalassaemia in which some α or β globin chains were synthesized, albeit at a reduced rate, no abnormalities in the patterns of synthesis were found. This suggested that the primary defect in these forms of the conditions must reflect a reduced amount of otherwise normal messenger RNA [15]. By the late 1960s, the field of protein chemistry and synthesis had taken the thalassaemia problem as far as it could go and further progress required the development of methods for studying mRNA levels, and later for defining the defects at the DNA level, work which proceeded rapidly from the early 1970s onwards. These experiments, and the later development of the field, are reviewed in detail elsewhere [16, 17].

Remarkable progress and discoveries in other aspects of the human haemoglobin field were made in Baltimore during this time, reflecting the interaction between different departments and disciplines which was encouraged by McKusick and Conley. One of the most important discoveries followed the finding of a family in which two haemoglobin variants segregated, one of which was a β globin variant and the other an α variant. This work left little doubt that there were at least two genetic loci involved in globin chain synthesis, the α and β loci, and, furthermore, that they are almost certainly on different chromosomes [18]. Conley's department also made a wide variety of observations on a condition which had been first noted in Africa, hereditary persistence of foetal haemoglobin, and clearly defined its genetic transmission, interaction with other haemoglobin variants and the first descriptions of its homozygous state [19, 20].

Another important contribution during this period was the discovery of an abnormal haemoglobin associated with hereditary polycythemia. This variant turned out to affect the α globin chain in the region of one of the critical α/β contacts which is involved in the allosteric changes consequent upon oxygenation of the haemoglobin molecule [21].

Conley's department also carried out a wide variety of studies on the pathophysiology of sickle cell anaemia and the interactions of the sickle cell gene, work which was later to culminate in the early trials of hydroxyurea for the management of the disease, the first genuinely successful approach to the control of sickle cell anaemia [22].

McKusick's Legacy

Those of us who trained with McKusick in the early 1960s were extremely fortunate. In particular, we were introduced to clinical genetics by a teacher who never lost touch with internal medicine and in an environment which was heavily embedded in general medicine; a similar situation would be extremely hard to find today.

As well as the achievements of the many fellows who passed through the Moore Clinic, McKusick's own work still touches on many aspects of the current field of medical genetics. He was undoubtedly one of the first to appreciate the importance of genetic linkage in clinical genetics, and his own group defined some of the first linkages of this type. His remarkable work on the Amish population clearly defined the importance of studying inbred groups as a unique opportunity for analysing and defining recessive disorders. His lifelong appreciation of the critical importance of genetic linkage was undoubtedly a major stimulus to developing a linkage map of the human genome and, later, he was one of the first proponents of the Human Genome Project. He was one of the first to appreciate the medical potential of this venture and was a founding member and the first president of the Human Genome Organization (HUGO). His remarkable book *Mendelian Inheritance in Man*, later provided online, became the major source of reference for clinical geneticists throughout the world. And his early association with the Jackson Laboratory, Bar Harbor, Maine, led to the initiation of the Summer Course in Medical Genetics which became an annual event which focused on updating teachers of the subject and which still flourishes and attracts a distinguished faculty.

McKusick's interests were not, of course, confined entirely to clinical genetics. He edited *The Principles and Practice of Medicine*, an updated version of Osler's Textbook of Medicine, and he also wrote widely in the field of medical history.

In short, McKusick's influence on the development of clinical genetics cannot be overemphasized. It is interesting to note that in his classical work, *Hereditary Disorders of Connective Tissue*, there is an illustrated dedication to Archibald Garrod. As mentioned earlier, Garrod and McKusick had a great deal in common in that they were practising physicians who saw the importance of genetics in medical practice and took their problems from the bedside to the laboratory and beyond. They were very remarkable men.

First McKusick lecture in April 2002. Victor MckUsick, C Lockard Conley, David Weatherall. Photograph taken on a visit to Conley after Weatherall had given the inaugural McKusick lecture

References

1. Bearn AG (1993) Archibald Garrod and the individuality of man. Clarendon, Oxford
2. Garrod AE (1909) Inborn errors of metabolism. In: Royal College of Physicians of London (ed) The Croonian lecture delivered before the Royal College of Physicians of London in June 1908. Frowde/Hodder and Stoughton, London
3. Garrod AE (1931) The inborn factors of disease. Oxford University Press, Oxford
4. McKusick VA (1956) Heritable disorders of connective tissue. Mosby, St. Louis
5. Weatherall DJ (1990) Presentation of the George M. Kober Medal to Victor A. McKusick. Trans Assoc Am Physicians 103:cxxii–cxl
6. McKusick VA (1966) Mendelian inheritance in man: catalogs of autosomal dominant, autosomal recessive and X-linked phenotypes. Johns Hopkins University Press, Baltimore
7. Weatherall D (2008) McKusick, Victor. In: Encyclopedia of life sciences. Wiley, Chichester
8. McKusick VA (1989) Forty years of medical genetics. JAMA 261:3155–3158
9. McKusick VA (2006) A 60-year tale of spots, maps, and genes. Annu Rev Genomics Hum Genet 7:1–27
10. Weatherall DJ, Boyer SH (1961) The genetic control of the α chains of human hemoglobins. Trans Assoc Am Physicians 74:89–99
11. Weatherall DJ, Boyer SH (1962) Evidence for the genetic identity of alpha chain determinants in hemoglobins A, A2 and F. Bull Johns Hopkins Hosp 110:8–22
12. Weatherall DJ, Baglioni C (1962) A fetal hemoglobin variant of unusual genetic interest. Blood 20:675–685
13. Weatherall DJ (1963) Abnormal haemoglobins in the neonatal period and their relationship to thalassaemia. Br J Haematol 9:265
14. Weatherall DJ, Clegg JB, Naughton MA (1965) Globin synthesis in thalassemia: an in vitro study. Nature 208:1061–1065
15. Clegg JB, Weatherall DJ, Na-Nakorn S, Wasi P (1968) Haemoglobin synthesis in β-thalassaemia. Nature 220:664–668
16. Weatherall DJ (2010) Thalassaemia. The biography. Oxford University Press, Oxford
17. Weatherall DJ, Clegg JB (2001) The thalassaemia syndromes. Blackwell Science, Oxford
18. Smith EW, Torbert JV (1958) Study of two abnormal hemoglobins with evidence for a new genetic locus for hemoglobin formation. Bull Johns Hopkins Hosp 102:38–45
19. Charache S, Clegg JB, Weatherall DJ (1976) The Negro variety of hereditary persistence of fetal haemoglobin is a mild form of thalassaemia. Brit J Haematol 34:527
20. Conley CL, Weatherall DJ, Richardson SN, Shepard MK, Charache S (1963) Hereditary persistence of fetal hemoglobin: a study of 79 affected persons in 15 Negro families in Baltimore. Blood 21:261
21. Charache S, Weatherall DJ, Clegg JB (1966) Polycythemia associated with a hemoglobinopathy. J Clin Invest 45:813
22. Charache S, Terrin ML, Moore RD, Dover GJ, Barton FB, Eckert SV, McMahon RP, Bonds DR (1995) Effect of hydroxyurea on the frequency of painful crises in sickle cell anemia. N Engl J Med 332:1317

Chapter 7
Cytogenetics and Early Days at the Moore Clinic with Victor McKusick

Malcolm A. Ferguson-Smith

Introduction

Victor McKusick's many achievements have rightly earned him the title of "founding father" of medical genetics. The trainees that he gathered around him at the Moore Clinic in 1957–1961 were his first medical genetics disciples, a number of whom went on to spread the word and create clinical genetics centers around the world. Victor's research was based on his clinical acumen and his ability to distinguish the signs and symptoms needed to define phenotypes and characterize specific syndromes. While laboratory diagnostics were part of genetic nosology, Victor was content to leave this aspect to others. His prodigious memory and boundless energy were put to good use in cataloging genetic disease, making teaching films, and in compiling chromosome maps. His organizational skills and fine teaching ability led to the renowned Bar Harbor course in genetics and to his crucial involvement in the Human Chromosome Mapping Workshops, in the foundation of HUGO, and in the establishment of the Human Genome Project.

Early Days at the Moore Clinic

Victor understood from the start that that progress in medical genetics depended, as in mouse and Drosophila genetics, on the availability of a human gene map. When in 1957 he was asked to take over the Moore Clinic and develop a new Division of Medical Genetics, the disciplines essential for gene mapping namely biochemistry, immunogenetics, and genetic linkage studies were soon put in place. The clinic was

M.A. Ferguson-Smith (✉)
Department of Veterinary Medicine, Cambridge Resource Centre for Comparative Genomics, Cambridge University, Madingley Road, Cambridge, CB3 0ES, UK
e-mail: maf12@cam.ac.uk

K.R. Dronamraju and C.A. Francomano (eds.), *Victor McKusick and the History of Medical Genetics*, DOI 10.1007/978-1-4614-1677-7_7,

named after Dr. J. Earle Moore who had established it as a famous center for the care of patients with chronic disease, and special sessions continued to be held for venereal disease, cancer, and other disorders. A number of Dr. Moore's trainees and fellows, including Tony Murphy, soon took up genetic studies under Victor's tutelage. Ned Boyer, who had joined Victor in 1956 as a fellow in cardiology, switched to biochemical genetics following a year's sabbatical with Harry Harris at the Galton in London. Murphy's expertise in statistics was channeled into population genetics. Immunogenetics was the province of Wilma Bias, and the responsibility for genetic linkage was taken by Jim Renwick. Renwick joined as a fellow in 1958 from the Galton and left early in 1959 to work with Guido Pontecorvo in Glasgow. He became a pioneer in devising computer programs for genetic linkage and collaborated closely with Victor and Jane Schultz, visiting the Moore Clinic for several weeks each year over the next 10 years to work with the Lockheed-Martin computer in Baltimore. David Price Evans from Liverpool and I were also recruited in 1958. Evans became an expert in pharmacogenetics, and I joined from Glasgow with the aim of being the team's cytogeneticist. David Weatherall, also from Liverpool, came in 1960 to work with Boyer and Lockhard Conley on the hemoglobinopathies, an area in which he became an international authority. I believe that Renwick, Evans, Jim Leeming, Ian Porter, and I, all British, were the first to be recruited as fellows into the new Division of Medical Genetics.

In 1959, the Division of Medical Genetics at Hopkins consisted of Victor, then associate professor of medicine, and at least 12 fellows in medicine, many inherited from Dr. Moore. Many nationalities were represented in the fellowship: Robin Bannerman (UK), Nemat Borhani (Iran), Dick Goodman (USA), Hymie Gordon (South Africa), Keith Manley (UK), Carmen Merriman (Cuba), Gururaj Mutalik (India), and Alan Skyring (Australia) plus others previously mentioned. The team included a dedicated group of research assistants (Margaret Hawkins, Barbara Latrobe), nurses (Nurse Hennessey), and secretarial staff (D. Parr, Jeanette Truscott, Ruth Kimmerer).

All the fellows attended the Genetics Clinic on Tuesdays which was followed by a conference, chaired by Victor, on the patients and families seen that day. Each fellow helped in the growth (cancer) clinic (Thursday, supervised by Al Owen) or venereal disease clinic (Friday, supervised by Dick Hahn). I was assigned to both, and this gave me my first experience of health services in the United States. We all had our specific research project and also helped in each other's projects. Thus, I was a subject in Evans' isoniazid polymorphism project [1] (I was a "slow" inactivator), and I helped Leeming and Merriman in measuring upper to lower segment ratios in African-American school children as controls for Victor's Marfan project [2]. Manley also worked on the isoniazid project, and Porter's project was on linkage between G6PD and color blindness [3]. Other projects at that time included elliptocytosis (Bannerman) [4], Buerger's disease (Goodman) [5], muscular dystrophy (Skyring) [6], familial atrial septal defect (Gordon) [7], and hypertension (Borhani).

On the evening of the first Wednesday in each month, all the fellows assembled at the McKusick home for a journal club. Each of us was responsible for scanning a number of medical journals for papers of genetic interest. We completed index cards for each item with full reference and summary of points of interest for presentation and discussion at the club. Victor made notes and collected the cards at the end. The meetings were informal, and the presentations were accompanied by coffee and large sticky buns provided by Anne McKusick. The work in preparing for the journal club was substantial but not wasted, as Victor used the index cards to help him compile annual reviews of genetics in medicine which were published as separate issues in the Journal of Chronic Diseases each year from 1958 to 1963 [8]. As most of the genetic papers cited were case histories of single gene defects, these reviews eventually evolved into what became the first contributions to Victor's Mendelian Inheritance in Man (MIM) [9]. He arranged them into catalogs of autosomal dominant, autosomal recessive, and X-linked traits. MIM was first published in 1966 and was regularly updated and edited by Victor. The printed version ran to 12 editions, the last in 1998, but an online version (OMIM) was available from 1987 and has continued ever since. OMIM remains an invaluable reference and is the most important database in clinical genetics and human genomics. It represents one of Victor's most remarkable achievements.

Cytogenetics at the Moore Clinic

Diagnostic cytogenetics came to the United States via Victor and his Division of Medical Genetics at Hopkins. I was fortunate to be accepted as a fellow in 1958 specifically to undertake research on human chromosomes. It seems appropriate to recount the circumstances if only as an appreciation of the part Victor played in making this possible.

At that time, I had been training in pathology in Glasgow under Dan Cappell and had been struggling for over a year to make chromosome preparations from patients with Klinefelter syndrome. Working on sex chromatin surveys with Bernard Lennox, we had discovered that cases of the syndrome were common among males with infertility and males with learning difficulties [10, 11]. It was thought at that time that males with sex chromatin were sex-reversed XX females, but this seemed to be refuted in one of our patients whose testicular biopsy revealed a solitary seminiferous tubule with complete spermatogenesis, including primary spermatocytes containing the XY sex bivalent [12]. As our Genetics Department was not prepared to make the necessary chromosome analysis, I was encouraged by Charles Ford at Harwell to try to do this myself. My work in pathology left little time to concentrate on this, and my attempts at bone marrow preparations [13] yielded only a few poor mitoses that, in my inexperience, I was unable to interpret. Dan Cappell had heard from one of his colleagues about Victor's new division and recommended me as a potential fellow in the hope that this would allow me the opportunity to develop my chromosome work. So, sometime in the summer of 1958, I was interviewed by Victor at the Adelphi Hotel in Liverpool. I explained to him my plan to determine whether or not there was a Y chromosome in Klinefelter syndrome. Victor kindly agreed to help and suggested that I should start my fellowship at Hopkins immediately. I was delighted with this and encouraged to learn that Harry Klinefelter was on the staff of the hospital. However, I postponed my departure until February 1959, in part because I wanted to complete a study on prepubertal Klinefelter cases. This work appeared in the Lancet [14] in January just before I sailed and, coincidentally, at the same time that the extra chromosome in Down syndrome was reported by Lejeune and Gauthier [15]. It seemed that I had arrived in the USA at a most opportune time for studying human chromosomes.

While the Moore Clinic was a wholly appropriate environment for clinical research, it lacked space for laboratory work and, in 1959, the only wet lab was a small biochemistry lab run by Boyer and his technician Linda Ulenum. I was assigned a small cupboard off the crowded secretarial office, without a window and with just enough space for one small table, a stool, and my microscope. To reach this space each day, I had to pass the desk of Marie Gzowska, a young college student who did part-time work for Victor and Abe Lilienfeld. She was detailed to help me with a buccal smear survey at Rosewood State Training School where we discovered patients with 47,XXX and 48,XXXY karyotypes. Marie and I were engaged later that year, and since then, we have enjoyed working together in cytogenetics and over 50 years of married life.

Victor arranged for me to be provided with a microscope and equipment for chromosome studies, and these arrived 4 weeks after I landed in Baltimore. However, I was not able to get bone marrow samples until some weeks later and so worked on material brought with me from Glasgow. My first fresh bone marrow proved unsuccessful, and then I had the good fortune to meet T.C. Hsu at a meeting of the American Tissue Culture Association that April in Atlantic City. Hsu invited me to visit his lab in Houston to learn some cytogenetics. It so happened that I had to give a talk at the Jefferson-Davies Hospital in Houston the following month, and so I took this opportunity to spend 3 days with T.C. Hsu and Albert Levan (then on sabbatical from Lund). Both were extremely helpful with advice on cell culture and chromosome technology, and Levan showed me how to draw chromosomes with a camera lucida. Within 2 weeks, and benefiting from their advice, I had obtained many excellent mitoses from bone marrow and was able to draw normal karyotypes with my new camera lucida. In the following weeks, I analyzed bone marrow samples from patients with Turner syndrome, male pseudohermaphroditism, and Down syndrome. Meanwhile, a 47,XXY karyotype was reported in a patient with Klinefelter syndrome by Patricia Jacobs [16] and a 45,X karyotype in Turner syndrome by Ford and Paul Polani [17]. Victor invited Levan to visit Hopkins later that summer, and I was able to show him my preparations on Down syndrome. Levan and I sat down with the camera lucida and determined that the extra small chromosome in my preparations was chromosome 22, the smallest acrocentric, and not chromosome 21 as Lejeune had claimed. As several other groups had now reported trisomy 21 in Down syndrome, we omitted to publish our result and missed the opportunity to correct a simple error in the accepted nomenclature which purported to number chromosomes in order of size. This is the reason that the smallest human acrocentric is still identified as chromosome 21 although everyone now accepts that chromosome 22 is the larger.

Space for making and analyzing chromosome preparations soon became a problem and, in the summer of 1959, permission was granted to convert the nearby men's

Fig. 7.1 Members of the cytogenetics team at the Moore Clinic in 1960–1961, including Rosemary Wise, Bonny Lewis, and Yvonne Hajda (*front row*); Stan Handmaker (*back row, left*); and Malcolm and Marie Ferguson-Smith (*back row, right*)

rest room and lavatory into a cytogenetics laboratory and photographic dark room, respectively. This worked well, and within a few months, we had an active production line busily making chromosome preps, photographing mitoses (camera lucida now supplanted by a modern microscope camera), developing and enlarging negatives, cutting out chromosomes from prints, and arranging them into karyotypes. With Victor's help, grants from the US Public Health Service (and later NIH) were obtained for equipment and the employment of research assistants. Yvonne Hajda, Bonnie Lewis, and Rosemary Wise (Fig. 7.1) were our first research assistants, and they were joined by students Stanley Handmaker, Dick Hill, Sandy Miller, and Lynn Ghant. Alan Johnston joined the team as a new fellow in 1960, followed by other fellows including Berwind Kaufmann, Peter Bowen, and Digamber Borgoankar. By this time, the bone marrow technique had been superseded by short-term cultures of peripheral blood [18], and the advantages of this in terms of sampling and methodology led to a rapid increase in referrals for diagnosis, both from the hospital and from much further afield. Thus, was our clinical cytogenetic diagnostic service born.

The peripheral blood method was not only convenient but also yielded many more mitotic cells, and the air-dried preparations were of far higher resolution than the marrow mitoses produced by the squash method. This provided the opportunity for improved chromosome identification and the correction of errors that had appeared in the literature. One such error concerned the identification of individual acrocentric chromosomes which was then based on the presence of satellites (which marked the sites of ribosomal DNA) on their short arms in only three of the five pairs. Indeed, the extra chromosome in Down syndrome was identified as 21 by its satellites that were thought to be absent in chromosome 22. It soon became apparent in our material that the five pairs were all satellited, and we were able to publish a normal karyotype showing satellites on nine of the ten acrocentrics [19]. We also noted a tendency of acrocentrics to attach to one another in mitoses by their satellites. We postulated that this "satellite association" was due to nucleolar fusion during interphase (or somatic pairing between homologous regions on nonhomologous chromosomes) and that this was a factor in the occurrence of chromosome aberrations by illegitimate recombination during meiosis, as two of the three autosomal trisomies and all known translocations at that time involved acrocentrics.

The main research interest of the cytogenetics team at the Moore Clinic was on sex chromosome abnormalities, and thus, the most numerous referrals were patients with Turner syndrome and pseudohermaphroditism from the pediatric endocrine clinic of Lawson Wilkins. True hermaphrodites and male pseudohermaphrodites were studied in collaboration with Howard W. Jones. These referrals led to publications on true hermaphroditism [20, 21], 48,XXXY Klinefelter syndrome [22], the triple-X syndrome [23], male pseudohermaphroditism [24], and Turner syndrome [25–28]. However, our earliest cytogenetic paper was with Bill Young and Tim Mertz and confirmed the diploid number of 48 in the chimpanzee using bone marrow [29]. The aim was to ensure that, in view of the recent revision of the human diploid number from 48 to 46, there had been no such mistake in the chimp.

The project on Turner syndrome involved chromosome analysis and clinical studies on 30 female patients who showed at least one of the characteristic features of the syndrome. It was our largest and most complex cytogenetic study during my period at the Moore Clinic. The results suggested for the first time that haploinsufficiency of genetic loci on the short arm of the X, homologous to genetic loci on the Y chromosome, were critical for the 45,X Turner phenotype [28]. A key point in our hypothesis was that the X-Y homologous loci escaped X-inactivation in normal females. At that time, this idea ran counter to Mary Lyon's conclusions on X-inactivation in the mouse, in which there was strong evidence that the phenotype of the XO mouse was no different from the normal XX mouse, i.e., that all X-linked loci were completely inactivated in one of the two X chromosomes in normal female mice [30]. Our results were thus controversial and, as a result, it proved difficult to convince reviewers and publication was delayed until 1964, although the studies had been completed in 1961. Case reports from the literature published after that date provided data in strong support of the haploinsufficiency hypothesis, and my review of the combined data in 1965 [31] became a citation classic in 1991.

The Role of Cytogenetics in the Development of Medical Genetics

I returned to Glasgow University in October 1961 and was appointed Lecture in Medical Genetics (based largely on my training under Victor) in Guido Pontecorvo's Department of Genetics. A small laboratory for human cytogenetics was soon set up with Marie Ferguson-Smith and Patricia Ellis as research assistants and Marion Dickson as our technician. We started work on chromosome polymorphisms and human nucleolar chromosomes in meiosis [32, 33]. Jim Renwick had moved to the same department in 1959 from the Moore Clinic to continue genetic linkage studies in families with genetic disorders and to establish a genetic marker laboratory headed by Marion Izatt. I also resumed my work on sex chromosomes by revisiting the Klinefelter cases that I had identified previously in nuclear sexing surveys among infertility patients and those in mental institutions. Over 120 Klinefelter patients were assembled and karyotyped over the next few years [34]. One of our aims was to study the parental origin of nondisjunction of the sex chromosomes using color vision tests and *XG* typing. Patricia Tippett at the laboratory of Rob Race and Ruth Sanger at the Lister Institute undertook the *XG* work while I visited the families with my color vision charts and took the blood samples. We found that maternal nondisjunction was associated with increased maternal age (as in Down syndrome) in comparison to the similar number of cases of paternal nondisjunction in which there was no such effect [32]. Among the patients that were *XG* typed were several with an apparently normal 46,XX female karyotype, referred to as XX males, and two of these had failed to inherit their father's *XG* (a) allele. It was suggested that this unexpected observation was most likely due to XX/XXY mosaicism. However, in a eureka moment, an alternative explanation occurred to me, namely that an illegitimate X-Y recombination during meiosis in the father could have resulted in the exchange of the male determinant on the Y for the *XG* locus on the X [35]. This proved to be correct, and the elusive male determinant, SRY, was isolated from such an XX male some 24 years later [36].

By 1962, our small research group was asked increasingly to provide chromosome analysis in patients with suspected chromosome abnormality, mostly referred from the Royal Hospital for Sick Children. The demand soon outstripped our limited resources, and this led in 1964 to the provision of a new laboratory for our diagnostic work located in the Queen Mother's Maternity Hospital, close to the Children's Hospital, and staffed by Marie Ferguson-Smith, Elizabeth Boyd, and Irene Greig. I was given a clinical appointment and started a weekly genetics clinic at the Queen Mother's Hospital. At first, most of the referrals were for chromosomal disorders, but metabolic and other more complex genetic diseases were soon encountered and, when Victor's Mendelian Inheritance in Man appeared in 1966, it proved to be an indispensible aid in my efforts at clinical diagnosis. Attendance at the clinic grew steadily, and after our paper in 1971 [37] on a prospective study of prenatal diagnosis for chromosome aberrations (the first in the UK), we had to expand the number of clinics. The risk of Down syndrome with increased maternal age was the most frequent indication for amniocentesis and, by 1981, we had tested 2,700 women aged 35 and over and detected 23 Down syndrome fetuses by chromosome analysis of amniotic cell cultures. John Yates and I collected amniocentesis data from European centers and calculated maternal age-specific rates for chromosome aberrations in nearly 53,000 couples in the second trimester [38]. These results were widely used for counseling. Our prenatal cases had the benefit of ultrasound guidance for inserting the amniocentesis needle thanks to Ian Donald, the pioneer of sonar in obstetrics, at the Queen Mother's Hospital. With ultrasound, gestation could be confirmed, twins identified, and the placenta located so that anterior placentas could be circumvented. The need for safe amniocentesis was an important factor in the developments in obstetric ultrasound that led to the visualization of fetal malformations.

In 1972, it was discovered by Brock and Sutcliffe that fetal neural tube defects (NTDs) could be detected in the second trimester by elevated levels of amniotic alpha-fetoprotein (AFP). Anencephaly and spina bifida were common malformations in the West of Scotland at that time, occurring in 5 per 1,000 births, and we were quick to make the first prenatal diagnosis of open spina bifida by this means in 1972 [39]. When it was realized that maternal serum AFP levels were also raised in affected pregnancies, we started a screening program for pregnant mothers in 1975 [40]. By 1982, we had screened over 140,000 pregnancies and detected 310 anencephalic and 195 open spina bifida fetuses [41]. The rate of NTDs in live and stillbirths was reduced from 4.3 per 1,000 in 1976 to 1.7 per 1,000 in 1981. By 1991, approximately 60% of pregnancies in Europe had maternal serum AFP screening for NTD, but since then, ultrasound has been used increasingly for both screening and diagnosis.

In 1984, it was found that maternal serum AFP levels were decreased in Down syndrome pregnancies, and this factor, plus other biochemical and ultrasound markers (together with maternal age), was exploited for screening for chromosomal trisomies. By 2009, screening protocols were improved sufficiently to detect 90% of Down syndrome pregnancies for a false-positive rate of 2% [42]. This reduced the need for amniocentesis and its albeit small risk of miscarriage. Recent work on the analysis of fetal nucleic acids in maternal blood has demonstrated that fetal sex and Rhesus genotype can be determined successfully in early pregnancy without the need for invasive procedures. Current research indicates that the diagnosis of fetal trisomies is also feasible by this method and should soon become available for patients [43].

The Human Gene Map and the Genome Project

Victor's interest in research on mapping human genes may have been prompted while he was engaged in compiling a paper on X-linked disorders [44]. Julia Bell and J.B.S. Haldane had described linkage between hemophilia and color blindness in 1937 [45], with improved estimates by Haldane and Cedric Smith in 1947 [46], but there were few similar studies reported at that time. Victor and Ian Porter exploited the recent observations on G6PD polymorphism among African-Americans in 1962 to test for linkage with color blindness and found 5% recombination [3]. This was the first genetic linkage study from the Moore Clinic. While autosomal linkage between genes for Lutheran and Lewis was discovered by Jan Mohr in 1951 [47], followed by linkage between the nail-patella syndrome and the ABO blood group loci by Renwick and Sylvia Lawler in 1955 [48], there were no chromosome assignments of human autosomal gene loci until the paper on linkage between the Duffy blood group and a centric chromosome 1 polymorphism by Roger Donahue, Bias, Renwick, and McKusick in 1968 [49]. The reason for the slow progress in human linkage studies was the dearth of common genetic polymorphisms. As the above examples demonstrate, the only available polymorphisms before the 1980s were blood groups, red cell enzymes, and a few serum proteins; in total, these amounted to only 23 autosomal loci for testing linkage with all known single gene defects. Most studies using chromosomal polymorphisms proved to be disappointing, with negative results for gene assignment. In 1980, a host of DNA polymorphisms (restriction fragment length polymorphisms) became available, and for the first time, the linkage approach became worthwhile.

Deletion mapping was an alternative approach that exploited the identification of unbalanced structural abnormalities detected during routine cytogenetic diagnosis. Our first success was in mapping the gene for red cell acid phosphatase to the end of the short arm of chromosome 2 in 1973 by loss of a parental allele in the deleted region and confirmed by enzyme dosage in the affected patient and his parents [50]. Other assignments soon followed from our group, including the loci for *AK1, GOT, NP, HPT, ADA, GALT*, and *XG*. The localization of *AK1* to chromosome 9q34 also mapped the closely linked *ABO*: nail-patella linkage group and the locus for tuberous sclerosis type 1 [51].

Meanwhile, the major strategy for gene mapping exploited somatic cell genetics and interspecific hybrids between human and mouse cells in which loss of human chromosomes allowed the assignment or exclusion of human proteins by biochemical assay of hybrids [52]. Victor and Frank Ruddle championed the somatic cell approach and organized the highly successful Human Gene Mapping (HGM) Workshops, the first of which was held by Ruddle in New Haven in 1973, and in which hybrid studies figured prominently. The 11th workshop was organized by Walter Bodmer in 1991. However, the accumulation of results by then made collating the data at one meeting an impossible task, and Single Chromosome Workshops took over, which I helped to organize under HUGO (London) and which continued until 1999. Throughout this period, Victor recorded, in tables and diagrams in successive editions of MIM, the progress being made in mapping human chromosomes. This was another valuable service greatly appreciated by cytogeneticists and all those of us involved in mapping.

Direct mapping of genes to human chromosomes under the microscope by in situ hybridization became possible in 1981, using RNA or cloned cDNA labeled first by tritium and then, about 10 years later, by fluorescent dyes. Earlier attempts to locate single-copy genes had been unsuccessful until DNA cloning allowed the production of improved DNA probes. We proved the feasibility of these new probes in 1981 by regional mapping the alpha- and beta-globin genes to the short arms of chromosomes 16 and 11, respectively [53]. We followed this by mapping the immunoglobulin kappa light chain gene to the proximal short arm region of chromosome 2 [54]. Since then, tritium labeling has been replaced by fluorochrome labeling. Fluorescence in situ hybridization (FISH) is now the standard first approach to assigning genes and DNA markers to chromosomes.

Accommodation for the West of Scotland Regional Genetics Service, which started in 1965, and the University Department of Medical Genetics, created in 1973, was always a problem until 1980 when we obtained our separate Institute to house both beside the Royal Hospital for Sick Children. This was made possible by funding from the National Fund for Research into Crippling Diseases, founded by Duncan Guthrie, first for polio research in the UK and later extended to include research into all handicapping disorders of children. We were delighted when Victor came to Glasgow in 1981 to open our building in the presence of Duncan Guthrie and other supporters from the fund, university, hospital, and region.

In 1987, I moved to head the Cambridge University Department of Pathology and to direct the East Anglia Regional Genetics Service at Addenbrooke's Hospital. Our research program encompassed the molecular pathology of genetic disease, and our gene mapping studies resulted in the positional cloning of disease genes including von Hippel-Lindau disease, multiple endocrine neoplasia, glycerol kinase deficiency, and hereditary persistence of fetal hemoglobin. Intensive efforts over 10 years to identify the gene for multiple self-healing epithelioma mutations have only recently born fruit with the discovery by next-generation sequencing of mutations in TGFBR1 [55]. FISH and dot blots of translocation derivatives sorted by flow cytometry were key strategies Nigel Carter and I developed for mapping at that time. Chromosome sorting followed by random-primed PCR labeling produced chromosome-specific DNA probes that we could use for the analysis of complex chromosome aberrations, a procedure we termed reverse chromosome painting [56]. We went on to produce multicolor FISH probes for both human and mouse by labeling each chromosome with a different combination of fluorochromes [57]. Chromosome aberrations could then be analyzed successfully by digital fluorescence microscopy in one M-FISH procedure.

Another very useful application of chromosome sorting was initiated by Fengtang Yang who joined our group in 1993. Yang was interested in studying karyotype evolution in muntjacs, and we were able to sort the three pairs of chromosomes in the Indian muntjac and label two of them with the two different fluorochromes that were available, and label the third with a mixture of the fluorochromes. Cross-species painting with other species of muntjac and deer revealed chromosome homologies and evolutionary rearrangements that have contributed to our understanding of cervid evolution [58]. Since then, cross-species painting has been a major occupation of our research group and our collaborators that has continued for the past 17 years. Yang, Willem Rens, and Patricia O'Brien have been the major contributors in our team, and we have studied representative species of all mammalian orders and have determined phylogenetic relationships including the putative ancestral eutherian karyotype [59]. Our studies resolved the nature of the ten sex chromosomes of the platypus [60] and its relationship to the echidna [61], and our cross-species painting now continues with projects on avian and reptilian species that beautifully illustrate Darwin's belief that all species descend from a common progenitor [62].

Fifty years ago, it was widely believed by biologists and geneticists that human cytogenetics was an outmoded discipline and not worthy of grant support.

Despite such bad omens, chromosome studies have flourished and have been major contributors to our knowledge of the human genome and to progress in medical genetics. Indeed, the genome sequence now defines all genetic disease. Details of the role of cytogenetics in the development of the human gene project, and of medical genetics in general, are reviewed elsewhere [63, 64, 65]. Victor's encouragement of human gene mapping and his influence with the March of Dimes, which led to their support of the HGM workshops, were critical factors in the generation of a comprehensive human gene map. It is clear that the Human Genome Project would have taken much longer to produce the draft DNA sequence of the human genome in 2001 had this map not been available, for it was the map that enabled the assembly of sequenced DNA fragments into their respective chromosomes and in the correct order.

Concluding Comment

The discovery of gross aberrations in human chromosomes in 1959 stimulated medical interest in genetics at a time when the Division of Medical Genetics had just been created. Victor McKusick was influential in establishing diagnostic cytogenetics at Johns Hopkins Hospital before such services were available elsewhere in the United States. He understood that chromosome maps of genes were vital for progress in medical genetics and for 50 years worked tirelessly to support the characterization and cataloging of single gene defects and the assignment of their loci to specific chromosomes. This contributed to the positional cloning of disease genes and, ultimately, to the discovery of causal mutations at the DNA level. In many cases, this revealed the probable function of normal genes. The ultimate aims were to improve the diagnosis, management, and prevention of genetic disorders, and these have been achieved for many conditions; only gene therapy remains elusive. Like many of his early disciples, I found my 3 years with Victor at the Moore Clinic to be hugely rewarding. The experience steered me into a career in medical genetics and cytogenetics during an exciting period during which these disciplines emerged from obscurity to become the sciences underpinning modern medical practice. Victor's contributions can be found in all aspects of this great advance.

Acknowledgment I am most grateful to Marie Ferguson-Smith for checking the manuscript and for recalling with me our happy and fruitful times with Victor and colleagues at the Moore Clinic.

References

1. Evans DAP, Manley KA, McKusick VA (1960) Genetic control of isoniazid metabolism in man. Br Med J 2:485–491
2. McKusick VA (1960) Heritable disorders of connective tissue, 2nd edn. Mosby, St. Louis

3. Porter IH, Schultze J, McKusick VA (1962) Linkage between the loci for glucose-6-phosphate dehydrogenase deficiency and color-blindness in American Negroes. Ann Hum Genet 26:107–122
4. Bannerman RM, Renwick JH (1962) The hereditary elliptocytoses: clinical and linkage data. Ann Hum Genet 26:23–38
5. McKusick VA, Harris WS, Ottesen OE, Goodman RM, Sheeley WM, Bloodwell RO (1962) Buerger's disease: a distinct clinical and pathologic entity. JAMA 181:5–12
6. Skyring AP, McKusick VA (1961) Clinical, genetic and electrocardiographic studies in childhood muscular dystrophy. Am J Med Sci 242:534–547
7. Bizarro RO, Callahan JA, Feldt RH, Kurland LT, Gordon H, Brandenburg RO (1970) Familial atrial septal defect with prolonged atrioventricular conduction: a syndrome showing the autosomal dominant pattern of inheritance. Circulation 41:677–684
8. McKusick VA et al (1964) Medical genetics 1963. J Chronic Dis 17:1077–1215
9. McKusick VA (1966) Mendelian inheritance in man: catalogs of autosomal dominant, autosomal recessive and X-linked phenotypes, 1st edn. Johns Hopkins University press, Baltimore
10. Ferguson-Smith MA, Lennox B, Mack WS, Stewart JSS (1957) Klinefelter's syndrome: frequency and testicular morphology in relation to nuclear sex. Lancet 2:167–169
11. Ferguson-Smith MA (1958) Chromatin-positive Klinefelter's syndrome (primary micro-orchidism) in a mental deficiency hospital. Lancet 1:928–931
12. Ferguson-Smith MA, Munro IB (1958) Spermatogenesis in the presence of female nuclear sex. Scott Med J 3:39–42
13. Ford CE, Jacobs PA, Lajtha LG (1958) Human somatic chromosomes. Nature 181:1565–1568
14. Ferguson-Smith MA (1959) The prepubertal testicular lesion in chromatin-positive Klinefelter's syndrome (primary micro-orchidism) as seen in mentally handicapped children. Lancet i:219–222
15. Lejeune J, Gautier M, Turpin R (1959) Les chromosomes somatique des enfants mongoliens. Compt Rend Acad Sci Paris 248:1721
16. Jacobs PA, Strong JA (1959) A case of human intersexuality having a possible XXY sex determining mechanism. Nature 183:302–303
17. Ford CE, Jones KW, Polani PE, de Almeida JC, Briggs JH (1959) A sex chromosome anomaly in a case of gonadal dysgenesis (Turner's syndrome). Lancet 1:711–713
18. Moorehead PS, Nowell PC, Mellman WJ, Battips DM, Hungerford DA (1960) Chromosome preparations of leucocytes cultured from human peripheral blood. Exp Cell Res 20:613–616
19. Ferguson-Smith MA, Handmaker SD (1961) Observations on the satellited human chromosomes. Lancet 1:638–640
20. Ferguson-Smith MA, Johnston AW, Weinberg AN (1960) The chromosome complement in true hermaphroditism. Lancet 2:126–123
21. Jones HW, Ferguson-Smith MA, Heller RH (1965) Pathological and cytogenetic findings in true hermaphroditism: report of six cases and review of 23 cases from the literature. Obstet Gynecol 25:435–477
22. Ferguson-Smith MA, Johnston AW, Handmaker SD (1960) Primary amentia and micro-orchidism associated with an XXXY sex chromosome constitution. Lancet 2:184–187
23. Johnston AW, Ferguson-Smith MA, Handmaker SD, Jones HW, Jones GS (1961) The triple-X syndrome: clinical, pathological and chromosomal studies in three mentally retarded cases. Br Med J 2:1046–1052
24. Alexander DS, Ferguson-Smith MA (1961) Chromosomal studies in some variants of male pseudohermaphroditism. Pediatrics 28:758–763
25. Tanner JM, Prader A, Habich H, Ferguson-Smith MA (1959) Genes on the Y chromosome influencing rate of maturation in man: skeletal age studies in children with Klinfelter's (XXY) and Turner's (XO) syndrome. Lancet 2:141–144
26. Boyer SH, Ferguson-Smith MA, Grumbach MM (1961) The lack of influence of parental age and birth order in the aetiology of nuclear-sex chromatin-negative Turner's syndrome. Ann Hum Genet 25:215–225
27. Jones HW, Ferguson-Smith MA, Heller RH (1963) The pathology and cytogenetics of gonadal agenesis. Am J Obstet Gynecol 87:578–598

28. Ferguson-Smith MA, Alexander DS, Bowen P, Goodman RM, Kaufmann BN, Jones HW, Heller RH (1964) Clinical and cytogenetical studies in female gonadal dysgenesis and their bearing on the cause of Turner's syndrome. Cytogenetics 3:353–383
29. Young WJ, Merz T, Ferguson-Smith MA, Johnston AW (1960) The chromosome number of the chimpanzee, *Pan trogolodytes*. Science 131:1672–1673
30. Lyon MF (1961) Gene action in the X-chromosome of the mouse (*Mus musculus* L.). Nature 190:372–373
31. Ferguson-Smith MA (1965) Karyotype-phenotype correlations in gonadal dysgenesis and their bearing on the pathogenesis of malformations. J Med Genet 2:142–155
32. Ferguson-Smith MA, Mack WS, Ellis PM, Dickson M, Sanger R, Race RR (1964) Parental age and the source of the X chromosomes in XXY Klinefelter's syndrome. Lancet 1:46
33. Ferguson-Smith MA (1964) The sites of nucleolus formation in human pachytene chromosomes. Cytogenetics 3:124–134
34. Ferguson-Smith MA (1966) Sex chromatin, Klinefelter's syndrome and mental deficiency. In: Moore KL (ed) The sex chromatin. Saunders, Philadelphia, pp 277–315
35. Ferguson-Smith MA (1966) X-Y chromosomal interchange in the aetiology of true hermaphroditism and of XX Klinefelter's syndrome. Lancet 2:475–476
36. Sinclair AH, Berta P, Palmer MS, Hawkins JR, Griffiths BL, Smith MJ, Foster JW, Frischauf AM, Lovell-Badge R, Goodfellow PN (1990) A gene from the human sex determining region encodes a protein with homology to a conserved DNA binding motif. Nature 346:240–244
37. Ferguson-Smith ME, Ferguson-Smith MA, Nevin NC, Stone M (1971) Chromosome analysis before birth and its value in genetic counselling. Br Med J 4:69–74
38. Ferguson-Smith MA, Yates JRW (1984) Maternal age specific rates for chromosome aberrations and factors influencing them: report of a collaborative European study on 52,965 amniocenteses. Prenat Diagn (Special Issue) 4:5–44
39. Allan LD, Ferguson-Smith MA, Donald I, Sweet E, Gibson AAM (1973) Amniotic fluid alphafetoprotein in the antenatal diagnosis of spina bifida. Lancet 2:522–525
40. Vince JD, McManus TJ, Ferguson-Smith MA, Ratcliffe JG (1975) A semi-automated serum alphafoprotein radioimmunoassay for prenatal spina bifida screening. Br J Obstet Gynaecol 82:718–727
41. Ferguson-Smith MA (1983) The reduction of anencephalic and spina bifida births by maternal serum alphafetoprotein screening. Br Med Bull 39:365–372
42. Cuckle HS, Malone FD, Wright D, Porter TF, Nyberg DA, Comstock CH, Saade GR, Berkowitz RL, Ferreira JC, Dugoff L, Craigo SD, Timor IE, Carr SR, Wolfe HM, D'Alton ME (2008) Contingent screening for Down syndrome – results from the FaSTER trial. Prenat Diagn 28:89–94
43. Lo YMD, Tsui NBY, Chiu RWK, Lau TK, Leung TN, Heung MMS, Gerovassili A, Jin Y, Nicolaides KH, Cantopr CR, Ding C (2007) Plasma placental RNA allelic ratio permits non-invasive prenatal chromosomal aneuploidy detection. Nat Med 13:218–223
44. McKusick VA (1962) On the X chromosome of man. Q Rev Biol 37:69–175
45. Bell J, Haldane JBS (1937) The linkage between the genes for colour blindness and haemophilia in man. Proc R Soc B 123:110–150
46. Haldane JBS, Smith CAB (1947) A new estimate of the linkage between the genes for color-blindness and hemophilia in man. Ann Eugen 14:10–31
47. Mohr J (1951) Estimation of linkage between the Lutheran and the Lewis blood groups. Acta Pathol Microbiol Scand 29:339–344
48. Renwick JH, Lawler SD (1955) Genetical linkage between the ABO and nail-patella loci. Ann Hum Genet 19:360–367
49. Donahue RP, Bias WB, Renwick JH, McKusick VA (1968) Probable assignment of the Duffy blood group locus to chromosome 1 in man. Proc Natl Acad Sci USA 61:949–955
50. Ferguson-Smith MA, Newman BF, Ellis PM, Thomson DMG, Riley ID (1973) Assignment by deletion of human red cell acid phosphatase gene locus to the short arm of chromosome 2. Nature 243:271–274
51. Connor JM, Pirrit LA, Yates JR, Fryer AE, Ferguson-Smith MA (1987) Linkage of the tuberous sclerosis locus to a DNA polymorphism detected by v-abl. J Med Genet 24:544–546

52. Weiss M, Green H (1967) Human-mouse hybrid cell lines containing partial complements of human chromosomes and functioning human genes. Proc Natl Acad Sci USA 58: 1104–1111
53. Malcolm S, Barton P, Murphy CST, Ferguson-Smith MA (1981) Chromosomal localisation of a single copy gene by in situ hybridisation: human beta-globin genes on the short arm of chromosome 11. Ann Hum Genet 45:135–141
54. Malcolm S, Barton P, Murphy C, Ferguson-Smith MA, Bentley DL, Rabbitts TH (1982) Localization of human immunoglobulin kappa light chain variable region genes to the short arm of chromosome 2 by in situ hybridization. Proc Natl Acad Sci USA 79:4957–4961
55. Goudie DR, D'Allessandro M, Merriman B, Lee H, Szeverenyi I, Avery S, O'Connor BD, Nelson SF, Coats ST, Stewart A, Christie L, Pichert G, Friedel J, Hates I, Burrows N, Whittaker S, Gerdes A, Broesby-Olsen S, Ferguson-Smith MA, Verma C, Lunny DF, Reversade B, Lane EB (2011) Multiple self-healing squamous epithelioma is caused by a disease-specific spectrum of mutations in the *TGFBR1* gene. Nat Genet 47:365–369
56. Carter NP, Ferguson-Smith MA, Perryman MT, Telenius H, Pelmear AH, Leversha MA, Glancy MT, Wood SL, Cook K, Dyson HM, Ferguson-Smith ME, Willatt LR (1992) Reverse chromosome painting: a method for the rapid analysis of aberrant chromosomes in clinical cytogenetics. J Med Genet 29:299–307
57. Schröck E, du Manoir S, Veldman T, Schoell B, Wienberg J, Ferguson-Smith MA, Ning Y, Ledbetter D, Bar-Am I, Soenksen D, Garini Y, Ried T (1996) Multicolor spectral karyotyping of human chromosomes. Science 273:494–497
58. Yang F, Carter NP, Shi L, Ferguson-Smith MA (1995) A comparative study of karyotypes of muntjacs by chromosome painting. Chromosoma 103:642–652
59. Ferguson-Smith MA, Trifonov V (2007) Mammalian karyotype evolution. Nat Rev Genet 8:950–962
60. Rens W, Grützner F, O'Brien PCM, Fairclough H, Graves JAM, Ferguson-Smith MA (2004) Resolution and evolution of the duck-billed platypus karyotype with an $X_1Y_1X_2Y_2X_3Y_3X_4Y_4X_5Y_5$ male sex chromosome constitution. Proc Natl Acad Sci USA 101:16257–16261
61. Rens W, O'Brien PCM, Grützner F, Clarke O, Graphodatskaya D et al (2007) The multiple sex chromosomes of platypus and echidna are not completely identical and several share homology with the avian Z. Genome Biol 8:R23
62. Pokorna M, Giovannotti M, Kratochvil L, Kasai F, Trifonov V, O'Brien PCM, Kratochvil L, Caputo V, Olmo E, Ferguson-Smith MA, Rens W (2011) Strong conservation of the bird Z chromosome in reptilian genomes is revealed by comparative painting despite 275 My divergence. Chromosoma 120:455–468
63. Ferguson-Smith MA (1993) From chromosome number to chromosome map: the contribution of human cytogenetics to genome mapping. In: Sumner AT, Chandley AC (eds) Chromosomes today, vol 11. Chapman & Hall, London, pp 3–19, Chapter 1
64. Ferguson-Smith MA (2008) Cytogenetics and the evolution of medical genetics. Genet Med 10:553–559
65. Ferguson-Smith MA (2011) Putting medical genetics into practice. Ann Rev Genomics Hum Genet 12:1–23

Chapter 8
Travels with Victor: Marfan Syndrome and Its Cousins

Reed E. Pyeritz

In the Beginning

My interest in genetics arose from research for my doctoral dissertation on repeated sequences in eukaryotic DNA [1]. My research mentor and I were invited to watch one of the first transabdominal amniocenteses conducted in Boston; he suffered vasovagal syncope, and I was hooked. By applying to medical school, my sole intent was to become a medical geneticist, certainly a novel career path in the early 1970s. I applied for residencies in both internal medicine and pediatrics, thinking either would prepare me clinically to be a geneticist. The luck of the draw had me at the Peter Bent Brigham Hospital in Boston as a medical intern. Each spring, our chief, noted cardiologist Eugene Braunwald, invited a senior physician-scientist from another academic medical center to serve for a week as the visiting professor. That is how, in 1976, I first met Victor McKusick. Dr. Braunwald knew Victor as a fellow cardiologist and physician in chief of a premier hospital. Since I knew of Victor's clinical studies on Marfan syndrome and had cared for a desperately ill young woman, pregnant and with an acute dissection of her entire aorta, earlier in the year, I was delighted by this choice of a visiting professor. During the week, Victor presented several conferences and the requisite medical grand rounds, along with leading morning report and attending rounds with several of the house staff teams. I was impressed, as anyone who ever met Victor was, with his breadth of knowledge. As a biochemist, I particularly related to his detailed discussions of the extracellular matrix and the recent discovery of a defect in type III collagen as the cause of type

R.E. Pyeritz, M.D., Ph.D. (✉)
Departments of Medicine and Genetics, Center for the Integration of Genetic Healthcare Technologies, The Perelman School of Medicine at the University of Pennsylvania, Philadelphia, PA 19104-4283, USA

Hospital of the University of Pennsylvania, Penn Tower 1115, 3400 Spruce St, Philadelphia, PA 19104, USA
e-mail: reed.pyeritz@uphs.upenn.edu

K.R. Dronamraju and C.A. Francomano (eds.), *Victor McKusick and the History of Medical Genetics*, DOI 10.1007/978-1-4614-1677-7_8,

IV Ehlers-Danlos syndrome (now termed the vascular form). He was legendary for having dozens of graduate students and postdoctoral fellows under his mentorship, and I envisioned many of them toiling away in his laboratory. Only later did I come to realize that Victor did not have a lab and never had.

Just before Victor's visit, I had accepted an invitation to join another institution as a fellow in medical genetics at the end of junior residency in medicine in mid-1977. I was already discussing the focus of an NIH-postdoctoral fellowship application. Knowing of my interest in genetics, Dr. Braunwald arranged for me to have a private meeting with the visiting professor before he returned to Baltimore. While I do not recall being put at ease in his presence—he was no more comfortable with small talk for more that 30 seconds then than he was later in life—the lasting impression was one of authority, both intellectual and personally. When I told him of my plans to pursue training in medical genetics elsewhere, he simply replied, "No, Reed, you need to come to Hopkins." More than slightly taken aback—and pleased to be so anointed—I promised to get back to him promptly with my decision. And that was that.

A follow-up letter to him was passed to Dr. Edmond A. Murphy, who was chief of the Division of Medical Genetics; "Tony" had taken over the division and the Moore Clinic when Victor became the Osler professor and chair of the Department of Medicine at Hopkins. Despite Victor's assurances of my position as, I thought, a fellow in medical genetics, the official invitation to join Hopkins was some months in arriving. This was in the days when the American Board of Internal Medicine granted board eligibility to those who completed 2 years of medicine residency followed by a postdoctoral fellowship. Consequently, I did not need to do a senior residency in medicine. My wife and I bought a house in Baltimore and were planning our move when I received a letter from David Robertson, a rising assistant chief of service (chief resident) of one of the four medical house staff firms. David welcomed me as a senior resident and asked me to select elective rotations for the year. A call to Victor's office eventually was patched through to him, and, in his characteristic hemming and hawing, a pattern of speech I was to hear for decades, he emphasized the great honor to be selected for the Osler Medical Residency, the wonderful opportunity to join the Barker Firm, and during that year, I could select a few elective months in medical genetics. By that point in my training, I was tired—literally and figuratively—of being on service every second or third night and was looking forward to focusing on genetics. But having burned all other bridges, I was committed to Hopkins at that point and resigned myself to serving what I viewed as an unnecessary year of senior residency. Only much later did I learn that this "opportunity" was the only way Victor could pay me since the funding for medical genetics fellows had dried up.

Life at Hopkins

Late in June 1977, my wife packed up our car and dog and headed for our new home. I was on call the night of June 30 in Boston and due to start my residency at Hopkins the next morning. Unshaven and wearing scrubs, I caught an early flight and took a cab to the Johns Hopkins Hospital. Not knowing what else to do, I went to Victor's office on the tenth floor of the Blalock building. His secretary, who was not expecting me, said that he was seeing a patient, and I could have a seat. Imagining Victor somewhere in a clinic, I was in danger of dozing when the door to his office opened and he emerged to retrieve a letter his secretary was typing. Seemingly nonplussed to see me, he said, "Come in, Reed, I want you to meet this family." And that was my formal introduction to Marfan syndrome at Hopkins.

Victor began academic life as a cardiologist and wrote a textbook on heart sounds, which included a technique he championed, phonocardiography. His interests in genetics were stimulated early in life by two occurrences: growing up on a dairy farm and observing the breeding of cows, and having an identical twin. In 1955, he melded his two interests in a seminal paper on Marfan syndrome [2]. Even though Marfan had described his case in 1896 [3], involvement of the aorta was not recognized until 1943 [4, 5]. The coauthor of one of those reports was a young pediatric cardiologist at Hopkins, Helen Taussig. By reviewing autopsy findings, both at Hopkins and the Baltimore morgue, Victor showed that the prevalence of cardiovascular disease was much higher in Marfan syndrome than was evident from the tools available at that time, auscultation and chest radiography. His paper in 1955 also first introduced the concept of heritable disorders of connective tissue, with Marfan syndrome a charter member.

When I arrived in Baltimore, in marked contrast to the 1960s, there were no genetics fellow and no full-time clinical geneticist in the Division of Medical Genetics. Dating from the independent development of pediatric genetics under Barton Childs and in the Moore Clinic under Victor in the 1960s, until 1992, genetic services at Hopkins were "balkanized" (Fig. 8.1). Thus, if I were to learn anything about genetic disorders in children, I figured I would have to take the initiative to attend pediatric genetics rounds. I soon realized that Victor's Medical Genetics Clinic, held each Thursday in the Moore Clinic on the second floor of the Carnegie building, welcomed patients of all ages. Granted, they tended to have one of the many disorders of short stature or heritable disorders of connective tissue. As a result, I became facile with relating to and examining children, long before I had any of my own. During senior residency on the Osler Medical Service, I was granted 4 months of service on the medical genetics rotation. Victor's secretary would call me when a consult was requested of "the chief," and after seeing the inpatient I would go to Blalock 10 to present my findings. Victor would then lead me back to the ward to see the patient. This process allowed me to observe a master clinician at the bedside, while introducing me to many of the secret passageways and shortcuts known only to someone who had been at the Johns Hopkins Hospital for 30 years.

The clinical activities of the Division of Medical Genetics revolved around a number of case managers who provided Victor with years of dedicated service. While several had nursing or social work backgrounds, some had no formal medical training. All, however, were focused on specific groups of conditions and knew the patients and families extremely well. This continuity of care was especially important to Victor's approach to clinical research. He urged patients to return, year after year, for several reasons. On the one hand, he and his fellows and the specialists

Fig. 8.1 The leaders of the first unification of Human and Medical Genetics at Hopkins in 1992. Standing on the *left*, Barton Childs, professor of pediatrics; and on the *right*, Victor McKusick, then university professor of Medical Genetics. Seated on the *left*, the author, clinical director of the newly formed Center for Medical Genetics; and on the *right*, Haig Kazazian, director of the center

who participated in the clinic would learn the natural history of often obscure conditions and become smarter in diagnosing and managing them. On the other hand, patients with rare disorders often had nowhere else to turn to receive medical care and support. The person in charge of the Marfan population was B. J. Latrobe. Over the years, I learned as much from her about my patients as I did talking to them myself. Given Victor's substantial administrative responsibilities running the Department of Medicine, which he assumed in 1973, the large population of people with Marfan syndrome was somewhat adrift. On that first day of my residency, seeing a father who had dissected his entire aorta and survived heroic surgery and his three children in Victor's office, I felt clinically and intellectually drawn to the condition. Sensing this, Victor gave me a copy of the 4th edition of *Heritable Disorders of Connective Tissue* [6], which had been published in 1972, and suggested I tackle some of the challenges the Marfan syndrome posed. During the months I was not scheduled for the medical genetics elective, to the consternation of my other service chiefs, I would still find time to see patients in the Thursday clinic, which B.J. gladly scheduled for me.

In preparation for my anticipated formal fellowship in medical genetics the following year, Tony Murphy encouraged me to apply for postdoctoral support. I was awarded postdoctoral grants from both the American Heart Association and the NIH. A few days before I was to start the fellowship, Tony informed me that I could not receive support from two grants for the same miniscule salary, so rather than return the money, Victor appointed me an instructor on the faculty (another example of his fiscal creativity). While my salary was still small, the division did not have to relinquish either grant, and my academic career was launched.

In the summer of 1978, I attended as a student the famous Short Course on Human Genetics (as it was called that year), organized jointly by Hopkins and the Jackson Laboratory and held in Bar Harbor, Maine [7]. If speed limits were only modestly exceeded, the drive from Baltimore took about 10 hours. I always kept a picture of Victor on the front seat since his monozygotic twin, Vincent, was chief justice of the Supreme Court of Maine. I figured any Maine state trooper who stopped me might be impressed that I had friends in high places. Reluctant to let the long drive be unproductive, I spent the time dictating what I had learned during my first year about Marfan syndrome into a tape recorder and had it transcribed into a rough manuscript. This eventually turned into a review article published in the *New England Journal of Medicine* in 1979, coauthored with Victor [8]. It remains one of my most highly cited papers and more than anything established to the world beyond Hopkins that I was a colleague of the man who was the preeminent authority on the condition.

One of the perquisites of being a faculty member in genetics at Hopkins was being asked to teach in Bar Harbor each summer. For many years, until I had children who complained, I continued to dictate a manuscript on the drive down the coast.

A Brief Retrospective on Marfan Syndrome

Between the years 1900 and 2000, life expectancy for the general population of the United States increased from 47.3 years to 76.8 years, an improvement of 62% [9]. The McKusick team conducted the first and best examination of life expectancy of Marfan syndrome, in 1972 [10]. They found that mean survival for males was in the 30s and for females was in the 40s, and more than 90% died of cardiovascular causes. The cumulative probability of survival was 48 years. The patients who comprised this cohort lived at a time when neither medical nor surgical intervention afforded any benefit to survival. By the mid-1990s, cumulative survival had increased to 72 years [11] and is undoubtedly approaching that of the general population today. This translates to a more than 60% improved survival in about four decades.

Improvements in medical and, especially, surgical management account for this dramatic achievement. As early as 1971, Victor was suggesting pharmacologic therapy, first with reserpine and then β-adrenergic blockade, to reduce stress on the proximal aorta [10]. With his support, my first NIH RO1 application (with Tony Murphy as P.I., because I lacked a faculty appointment) proposed a randomized trial of propranolol, the only β-blocker then on the American market. This was funded and the clinical study was started, randomizing patients to either nothing (then the standard approach) or propranolol in escalating doses tailored to reduce both chronotropy and inotropy. If I, or Victor or Tony, had known then that the study would occupy more than a decade, with no guarantee of ongoing funding or my employment at Hopkins, we might have thought better of enrolling the first subject. Fortunately, for both the trial and my career, I was able to stick around, and the results, published in 1994, demonstrated a clear benefit in reduction in dissection and need for surgery in the treated group [12]. This study was only made possible by the introduction in the early 1970s of transthoracic echocardiography, which enabled the noninvasive monitoring of aortic root dimensions. The prophylactic use of β-blockade has been confirmed in a number of additional studies, some randomized and some utilizing historical controls [13, 14]. Currently, a novel approach to therapy, reducing the activity to transforming growth factor-β by means of an angiotensin receptor blocker, shows promise in both mice and a few children with severe features [15, 16]. Losartan is being tested formally against β-blockade, but the results are likely several years off [17].

The chief of cardiac surgery at Hopkins when I started was Vincent Gott, one of the kindest, most engaging, and innovative surgeons I had, and have, ever met. Vince was a pioneer in use of cardiopulmonary bypass and valve replacement before coming to Hopkins. But even he was reluctant to operate on people with Marfan syndrome in the mid-1970s because the published outcomes were so dismal (40–60% hospital mortality). Indeed, several of his cases from that era, reluctantly taken to surgery because of acute dissection, died in the operating room. He learned of a new technique for replacing the dilated aortic root that involved a composite graft, a woven conduit with a prosthetic valve sewn in one end. This approach had been developed by a London surgeon, Mr. Hugh Bental, on the spur of the moment as a Marfan patient was exsanguinating in the operating theater [18]. That patient lived, and the brief case report, published in 1969, took a few years to travel across the Atlantic. Even then, it was assumed that the larger the diameter of the aortic root aneurysm, the greater the risk of dissection and rupture (based on the LaPlace relationship). But the critical diameter was uncertain, and we suspected it would vary from one patient to another. With more than a little trepidation, Vince agreed to repair electively the ascending aortas in several of our Marfan patients who had large aneurysms, even though they had not dissected and had only moderate or less aortic regurgitation. They did well and thus began the seminal transition to prophylactic surgery. The initial Hopkins surgical series was published in 1986, and among 50 elective procedures, there was no hospital mortality, a truly stunning achievement [19]. The average aortic root diameter in this cohort of patients was 71 mm. As experience in the hands of Gott and talented surgeons elsewhere (notably, Hans Borst in Hannover, Stanley Crawford in Houston, James Kirkland in Birmingham, Nicholas Kouchoukos in St. Louis, Jack Matloff at Cedars Sinai in Los Angeles,

Norman Shumway at Stanford, and Magdi Yacoub in London) grew and technical nuances evolved, the diameter of the aortic root at which elective surgery was considered fell to 60 mm, then 50 mm [20, 21]. This undoubtedly saved many lives because the surgeons were largely unaware of the patients who died while their physicians waited for their aneurysms to reach 60 mm.

The past decade has seen the emergence of valve-sparing aortic root replacement, which serves several purposes. First, the native aortic valve is buttressed and retained, sparing the patient lifelong anticoagulation. Second, the operation can be performed on even smaller aneurysms; I now refer patients for surgery when the root dimension reaches 45 mm, especially if there is a family history of dissection. The jury is still out on the long-term reliability of this technique, but a new generation of gifted surgeons (Tirone David in Toronto, Craig Miller at Stanford, Joseph Coselli in Houston, Duke Cameron at Hopkins, Joseph Bavaria at Penn, to name a few), following in their mentor's large footsteps, is committed to studying the effectiveness of valve-sparing root repair and improving it when possible [22–24]. The clinical history of patients who underwent a "remodeling" operation and subsequently developed dilatation of retained sinus tissue taught all of us about the inadequacies of this technique. The "reimplantation" or "David V" approach obviates this complication and is used, with local modifications, exclusively in Marfan syndrome today.

The National Marfan Foundation

Soon after I arrived at Hopkins, Victor asked me to attend the 1977 annual meeting of the Little People of America (LPA) in Portland, Oregon. Kay Smith, the case manager for people of short stature in the Moore Clinic, each year raised enough funds from a secret list of donors to support the travel of several medical professionals to the LPA meeting. One of those donors was Alan C. (Ace) Greenberg, CEO of Bear Stearns long before it disappeared in the financial crisis of 2007–2008. Eventually, Mr. Greenberg gave major gifts to endow both the skeletal dysplasia clinic in what became the McKusick-Nathans Institute of Genetic Medicine and the McKusick Chair in Genetics and Medicine. Attending the LPA meeting was eye-opening for me, to say the least. First, I knew nothing of the variety, causes, complications, and management of the dozens of forms of short stature I saw on arrival. Second, I quickly learned to sit down when talking to a little person. Third, I made friends with colleagues from around the country, some of whom have authored essays in this volume, who patiently instructed me in the medical, cultural, and psychosocial nuances of short stature. Finally, I was amazed by the energy of the attendees, many with families, and the obvious support and enjoyment that derived from the opportunity to interact with each other.

When I returned to working with my growing clinic of people with Marfan syndrome, and developing an interest in achondroplasia (the "long and the short of it" I often quip), I became convinced that an LPA-like organization was needed for the tall folks. It took a while for the plans to germinate in my mind, but by early 1979, I broached the idea of a Marfan support group to Victor and, most importantly, B. J. Latrobe. Both were highly supportive, so I introduced the idea to a couple dozen of my patients who lived in the Baltimore-Washington corridor. Most of them were also enthusiastic, and I planned a gathering to discuss how to proceed. A meeting in a conference room at Hopkins would invariably be cold and impersonal. Since I lived in northern Baltimore County, I worried that distance would be an impediment to hosting the meeting. Victor rose to the occasion and volunteered his spacious home in Guilford, a residential neighborhood within the city limits where many Hopkins faculty lived. By any account, the Sunday afternoon meeting was a success, and we all left with the mission to gestate a support group. A full accounting of the fits and starts of incubation, birth, and incorporation warrants a fuller historical reckoning elsewhere. Suffice it to say that my original aspirations of having an organization devoted to educating about MFS, providing support to people with MFS, and supporting research about MFS and related conditions have been fully realized [www.marfan.org]. As the official incorporation of the National Marfan Foundation (NMF) occurred in 1981, 2011 marked its 30th anniversary. The first few annual national meetings were held on the Hopkins medical campus and included a clinic to enable those who could not benefit from clinical expertise near their homes to be evaluated by specialists well versed in Marfan syndrome from Hopkins and elsewhere. Such a clinic remains an integral component of the annual meetings today. The NMF established the Antoine Marfan Award as its premier recognition of a health professional who had made groundbreaking contributions to the field. At an early national meeting, held in the Turner Auditorium where he had presided over so many medical grand rounds, Victor was honored with the first Antoine Marfan Award.

In 1989, Victor and I hosted the first international research symposium on MFS in Baltimore in conjunction with an NMF meeting [25]. These symposia have been held subsequently at regular intervals, with the eighth occurring in 2010 [26].

Lumping and Splitting

For those interested in clinical genetics, certainly in the early days but even now, a common conundrum was whether people with similar, but perhaps not identical, phenotypes had the same or distinct conditions. In genetic jargon, should phenotypes be lumped or split? The emergence in the 1960s and 1970s of genetic testing, especially by cytogenetic and biochemical means, rendered this an ever more interesting challenge. In this regard, Victor was a pragmatist. Like most academicians, he was interested in splitting, when the phenotypic (including radiologic), inheritance, or laboratory studies showed clear distinctions among patients. He did this to great effect in disorders of short stature and the mucopolysaccharidoses. The situation with Marfan syndrome was a bit more fluid and remained so for years. In the 1960s, Victor had a notable success in splitting by practicing what he termed

Fig. 8.2 The team that discovered the genetic basis for Marfan syndrome in 1991 [33]. From *left*, Hal Dietz, Victor McKusick, Clair A. Francomano, Garry Cutting, and Reed Pyeritz

"mail-order genetics." He wrote to ophthalmologists around the country asking them to send a urine specimen on their Marfan patients with dislocated lenses. Based on a series of patients in Northern Ireland with a newly described condition, homocystinuria [27], Neil Schimke, then a fellow, analyzed the urine with the cyanide-nitroprusside test and found reducing substances in some specimens [28]. This confirmed that some people diagnosed with Marfan syndrome actually had homocystinuria, a great example of splitting. Another example involved delineation of congenital contractural arachnodactyly, which, in the view of some, [29] affected the young girl originally described by Marfan [3]!

Lumping was another matter. He had accumulated a large series of patients, typically with aortic aneurysms, whom he diagnosed as having "Marfan syndrome," and managed them accordingly. I inherited many of these folks, and while some of them were very different from the "classic" phenotype that he portrayed in the 4th edition of *Heritable Disorders of Connective Tissue*, who was I to disagree with the master's diagnosis? Things began to change in 1986 as a series of self-appointed expert panels promulgated diagnostic criteria for Marfan syndrome [30–33]. Simply based on phenotype, patients with bicuspid aortic valve syndrome, MASS phenotype, and familial aortic aneurysm and dissection in the absence of many other features of Marfan syndrome were teased out [34]. The discovery in 1991 (Fig. 8.2) that defects in the fibrillin-1 extracellular matrix protein caused MFS hastened splitting a bit, but not as much as many expected [35]. But there was relatively little harm done by continued lumping, since patients managed as if they had MFS, whatever their

Table 8.1 Conditions with features that overlap with Marfan syndrome

Condition	Gene(s)	OMIM number(s)
Arterial tortuosity syndrome	*SLC2A10*	208050
Bicuspid aortic valve	? (rarely *NOTCH1*)	109730
Congenital contractural arachnodactyly	*FBN2*	121050
Ehlers-Danlos syndrome, classic type	*COL5A1*, *COL5A2*	130000
Ehlers-Danlos syndrome, vascular type	*COL3A1*	130050
Familial aortic aneurysm and dissection	*ACTA2*, *MYH11*, *MYLK*	132900, 607086, 607087
	TGFBR1, *TGFBR2*	611788, 613780
Familial ectopia lentis	*FBN1*, *ADAMTSL4*	129600, 225100
Homocystinuria	*CBS*	236200
Loeys-Dietz syndrome	*TGFBR1*, *TGFBR2*, *SMAD3*	609192, 610168, 613795
MASS phenotype	? (rarely *FBN1*)	604308
Shprintzen-Goldberg syndrome	? (rarely *FBN1*)	182212
Stickler syndrome	*COL2A1*, *COL9A1*, *COL11A1*, *COL11A2*	108300, 614134, 604841, 184840
Weill-Marchesani syndrome	*ADAMTS10*, *FBN1*	277600, 608328

actual diagnosis, did quite well in general. This was true even of the couple of young patients who I followed as MFS, but who had hypertelorism and no lens dislocation, and later proved to have Loeys-Dietz syndrome (LDS) [36]. We now know that some patients with the full LDS phenotype need even more aggressive prophylactic aortic root surgery [37]. Other patients lumped as Marfan syndrome by me had aneurysms and dissections of arterial branches of the aorta and even coronary artery disease. Through the pioneering work of Dianna Milewicz, we now recognized that patients in these autosomal dominant families often have mutations in genes encoding vascular smooth muscle cytoskeletal proteins, such as *ACTA2* [38, 39]. Using the Marfan standards for repairing aortic root aneurysms works well even today, but as with the LDS patients, extensive vascular imaging by MRA or CTA at regular intervals is much more important. One source of satisfaction over the years is the realization that we indeed do get smarter [40, 41]. Table 8.1 lists a baker's dozen conditions with features that overlap with classic MFS and are often in the differential diagnosis when a patient is first seen in consultation. The discovery of the molecular bases of many of these conditions often permits precise diagnosis. However, while mutations in *FBN1* account for more than 95% of all people with classic MFS, mutations in this gene also cause familial ectopia lentis, MASS phenotype, familial aortic aneurysm, and, remarkably, Weill-Marchesani syndrome in which features are often the mirror image of MFS.

Natural History and Clinical History

By scheduling patients with rare disorders to return on a regular basis to the Moore Clinic, Victor enabled himself and all of his acolytes to document *natural history*, that is, what happens to the patient in the absence of effective (or ineffective but disruptive) attempts at management. Eventually, medical and surgical management (e.g., beta-blockade in MFS and wide lumbar laminectomy in achondroplasia) altered natural history, and the imperative to learn *clinical history* gained in importance. Victor encouraged me to follow my Marfan patients with these objectives in mind, and I had no hesitation given how challenging, delightful, and thankful almost all of them were. I orchestrated an important transition in the Moore Clinic in the early 1980s by recruiting Barbara Bernhardt away from Neil Schimke in Buffalo. Barbara was a card-carrying genetic counselor, one of the first at Hopkins. Not only did she bring rigor to the counseling of patients with many disorders, but she became an avid colleague in research. Over three decades, she and I have conducted research on a wide array of issues—clinical, ethical, and social [42–44]. Of the many possible examples of descriptive clinical research conducted over my 16 years at Hopkins and subsequently, in addition to those mentioned above, several stand out.

Pregnancy

From the literature through the 1970s, it seemed clear that women with MFS often died during or shortly after pregnancy, and therefore the counseling virtually all women of childbearing age received was to avoid conception. A number of women with MFS, including patients of mine in those early days, tearfully accepted that recommendation and underwent tubal ligation. However, clearly not all women with MFS died (ascending aortic dissection was and is the major concern) during pregnancy. I reviewed both the literature and all of the Hopkins cases in 1980 and documented two useful points [45]. First, the women who died of dissection almost always had preexisting evidence of substantial aortic root enlargement. Second, the many women who had uncomplicated pregnancies and deliveries had mild or no aortic dilatation. This led to a recommendation, which I made with trepidation to my patients and colleagues for more than a decade, that if a young woman's aorta measured 40 mm or less, pregnancy was relatively safe. The risk was certainly higher than the general population, and I encouraged management in a high-risk obstetric setting. Fortunately, my patients experienced very few complications. My obstetric colleagues and I conducted a prospective study during the early 1990s that involved regular transthoracic and abdominal ultrasound, and we documented virtually no progressive aortic dilatation in women who started with minimal dilatation of the aortic root [46]. One woman suffered a type B dissection, and there are a few similar case reports of this over the past two decades. But for now, the 40-mm "rule" still pertains in my clinic.

Dural Ectasia

There were several case reports of lumbar dural (arachnoid or Tarlov) cysts in the Marfan syndrome, but no systematic survey had been conducted. Because we were following so many post-cardiac surgery patients with computed tomography, I used the availability of the images to assess prevalence and severity of lumbosacral involvement. The results were impressive; about two-thirds of adult Marfan patients had some degree of erosion of the vertebral bodies, and many had enlargement of the neural foramina and frank anterior meningoceles [47]. Such changes are quite uncommon in any other conditions (except LDS), and for the second iteration of the international diagnostic criteria [32], dural ectasia was a major criterion. However, MR or CT imaging is required to identify the typical changes, and many centers around the world do not have ready access to such imaging, and in the United States, the procedure might not be reimbursed if it were ordered for diagnostic (not symptomatic) purposes. Thus, in the most recent version of the diagnostic criteria [33], dural ectasia is relegated to one of many possible "systemic" features.

The natural history of dural ectasia in Marfan syndrome remains to be determined. Clearly, some people experience pain, neurologic signs, or both [48]. Surgical therapy is generally ineffective and potentially hazardous. As most people with Marfan syndrome age to years never before experienced, dural ectasia may for some be a major determinant of quality of life.

Pectus Excavatum

Most people with Marfan syndrome have some deformity of the anterior chest. The pathogenesis is thought to be overgrowth of the tubular bones, the ribs. Pectus excavatum, often asymmetric, is the most common deformity. Most lesions are of cosmetic concern (not to minimize the importance of that issue), but some are deep enough to impair venous return (especially with exertion) and restrict vital capacity of the lungs. The thoracic surgeon, Mark Ravitch, introduced a procedure in 1949 in which the sternum is disarticulated from the ribs, repaired, and supported during healing with a substernal, stainless steel bar [49]. The bar is removed via a simple skin incision 6 months or so after the costal cartilages have regrown. This and other surgical techniques were being widely applied in children with Marfan syndrome, but by the time of adolescence, the defect was often as severe as previously. Alex Haller was the preeminent pediatric thoracic surgeon at Hopkins and had witnessed some of the recurrences. We collaborated on retrospective and prospective observations that led to the firm conclusion that surgery should be delayed in Marfan syndrome until the patient was close to skeletal maturity [50, 51]. Now it is time for someone to evaluate the more recent, less invasive Nuss procedure [52] in Marfan patients.

Pulmonary Function

A common complaint of young Marfan patients is lack of physical endurance or becoming easily fatigued on moderate exercise. Attention is often drawn to lung function, and it became clear to me early on that standard pulmonary function studies virtually always produce abnormal results, typically reduced forced vital capacity and total lung volume. One of our fellows, Elizabeth Streeten, and the pulmonologists rose to the occasion. We showed convincingly that usually the results were spurious [53]. Standard pulmonary function tests are interpreted in relation to several patient characteristics, including smoking status, age, gender, and *height*. Of course, people with Marfan syndrome are abnormally tall due to their long legs, which should not affect lung volumes. We showed that a better variable is the sitting height. When this measurement was utilized, only patients with important degrees of scoliosis and pectus excavatum had truly reduced lung volumes.

This insight pertains equally to some disorders of short stature. At an LPA convention in Boston, we were able to study the pulmonary function of people with achondroplasia and showed that, because of their relatively normal trunk dimensions and shortened limbs, sitting height was a better predictor of lung function and volumes [54].

Visceral Cysts

One of the first patients to have composite graft repair of the ascending aorta performed by Vince Gott was a young woman who I follow to this day. We have imaged her aorta up and down dozens of times over the decades. The reports of these studies invariably also reported a supposedly incidental finding of a large cyst atop her left kidney. This cyst became quite large, but never caused any clinical problem. In her images, and in those of many other patients, I began to notice other cysts, usually small, in the liver and the kidney. Cysts, particularly renal ones, are common as humans age and are usually benign and harmless. But I wondered if we were seeing something distinct in the Marfan syndrome. One of my radiology colleagues at Penn, Harold Litt, was cajoled into looking at this systematically, and we showed that indeed visceral cysts are part of the Marfan syndrome [55]. The pathophysiology is unclear, and fortunately, no one (to my knowledge) has experienced a complication. But knowledge of this age-related, pleiotropic feature is important so that well-meaning radiologists and other physicians do not start recommending and performing biopsies.

Dénouement

Perhaps the most satisfying aspect of this personal history with Marfan syndrome and Victor McKusick that started in 1977 is the following. A number of children and adolescents who I diagnosed and managed in the early years have gone on to have great lives. Moreover, they have had their own children, and I have been privileged to reassure them when an offspring clearly was unaffected and reassure them that the future was optimistic when I made the diagnosis of Marfan syndrome. Some of those offspring are now off to college, and, if I am lucky, I may yet see a third generation.

At the 2011 annual meeting of the National Marfan Foundation in Portland, Oregon, I was honored with the Founders Award. It was both a surprise and an emotional moment for me, but in accepting the award, I held it together long enough to note that none of what I and the NMF had accomplished would have been possible without Victor McKusick.

Travels with Victor officially ended with his death in 2008. I likely echo the sentiments of many of the contributors to this volume in believing and appreciating that Victor's influence will persist until my own academic travels end. I and several other contributors to this volume had the opportunity to compose obituaries about Victor in 2008, so I will not reiterate what I reminisced then [56].

Acknowledgments I thank all of the colleagues, both at Hopkins and elsewhere, who tolerated my entreaties to study one aspect or another of the patients we saw jointly. Particular thanks are due to Barbara Bernhardt, M.S.; Hal Dietz, M.D.; Vincent Gott, M.D.; Alex Haller, M.D.; Karen Hofman, M.D.; the late Steve Levin, D.D.S., M.S.; Irene Maumenee, M.D.; Judith Rossiter, M.D.; Paul Sponseller, M.D.; Thomas Traill, M.D.; and Kenneth Zahka, M.D. I am also grateful for the patients and relatives who taught me so much about the Marfan syndrome, collaborated on many laboratory and clinical research projects, and were instrumental in forming the National Marfan Foundation. I am especially moved by the contributions of the late Faith Clark, Ph.D.; the late Cheryll Gasner, R.N.; the late MaryAnn Roney, R.N.; the late Julie Conant, M.D.; the late Julie Kurnitz; Wendy Weiss; and especially Priscilla Ciccariello. I appreciate helpful comments on this essay by Barbara Bernhardt, Jacqueline Hecht, and Jane Tumpson. The author is a senior fellow of the Center for Bioethics and a senior fellow of the Leonard Davis Institute of Health Economics of the University of Pennsylvania. Preparation of this essay was supported by a grant from the National Human Genome Research Institute (5P50HG00487) and occurred while in residence at the Brocher Foundation, Hermance, Switzerland.

References

1. Pyeritz RE, Thomas CA Jr (1973) Regional organization of eukaryotic DNA sequences as studied by the formation of folded rings. J Mol Biol 77:57–74
2. McKusick VA (1955) The cardiovascular aspects of Marfan's syndrome: a heritable disorder of connective tissue. Circulation 11:321–341
3. Marfan AB (1986) Un cas de déformation congènital des quatre membres plus pronouncée aux extrémitiés charactérisée par l'allongement des os avec un certain degré d'amonassesment. Bull Mem Soc Med Hop Paris 13:220–226
4. Baer RW, Taussing HB, Oppenheimer EH (1943) Congenital aneurysmal dilatation of the aorta associated with arachnodactyly. Bull Johns Hopkins Hosp 72:309–331
5. Etter LE, Glover LP (1943) Arachnodactyly complicated by dislocated lens and death from rupture of dissecting aneurysm of the aorta. JAMA 123:88–89
6. McKusick VA (1972) Heritable disorders of connective tissue, 4th edn. Mosby, St. Louis
7. McKusick VA, Naggert J, Nishina P, Valle D (1999) 40 years of the annual 'Bar Harbor Course' (1960–1999): a pictorial history. Clin Genet 55:398–514
8. Pyeritz RE, McKusick VA (1979) The Marfan syndrome—diagnosis and management. N Engl J Med 300:772–777

9. Centers for Disease Control and Prevention (2011) Ten great public health achievements-worldwide, 2001–2010. MMWR Morb Mortal Wkly Rep 60:814–818
10. Murdoch JL, Walker BA, Halpern BL, Kuzma JW, McKusick VA (1972) Life expectancy and causes of death in the Marfan syndrome. N Engl J Med 286:804–808
11. Silverman DI, Burton KJ, Gray J, Bosner MS, Kouchoukos NT, Roman MJ, Boxer M, Devereux RB, Tsipouras P (1995) Life expectancy in the Marfan syndrome. Am J Cardiol 75:157–160
12. Shores J, Berger KR, Murphy EA, Pyeritz RE (1994) Chronic ß-adrenergic blockade protects the aorta in the Marfan syndrome: a prospective, randomized trial of propranolol. N Engl J Med 330:1335–1341
13. Salim MA, Alpert BS, Ward JC, Pyeritz RE (1994) Effect of beta-adrenergic blockade on aortic root rate of dilation in the Marfan syndrome. Am J Cardiol 74:629–633
14. Ladouceur M, Fermanian C, Lupoglazoff JM, Edouard T, Dulac Y, Acar P, Magnier S, Jondeau G (2007) Effect of beta-blockade on ascending aortic dilatation in children with Marfan syndrome. Am J Cardiol 99:406–409
15. Brooke BS, Habashi JP, Judge DP, Patel N, Loeys B, Dietz HC (2008) Angiotensin II blockade and aortic-root dilatation in Marfan's syndrome. N Engl J Med 358:2787–2795
16. Pyeritz RE (2008) A small-molecule for a large disease (editorial). N Engl J Med 358: 2829–2831
17. Lacro RV, Dietz HC, Wruck LM, Bradley TJ, Colan SD, Devereux RB, Klein GL, Li JS, Minich LL, Paridon SM, Pearson GD, Printz BF, Pyeritz RE, Radojewski E, Roman MJ, Saul JP, Stylianou MP, Mahony L (2007) Rationale and design of a randomized clinical trial of beta blocker therapy (atenolol) vs. angiotensin II receptor blocker therapy (losartan) in individuals with Marfan syndrome. Am Heart J 154:624–631
18. Bentall H, De Bono A (1969) A technique for complete replacement of the ascending aorta. Thorax 23:338–339
19. Gott VL, Pyeritz RE, Magovern GJ Jr, Cameron DE, McKusick VA (1986) Surgical treatment of aneurysm of the ascending aorta in the Marfan syndrome: results of composite-graft repair in 50 patients. N Engl J Med 314:1070–1074
20. Gott VL, Greene PS, Alejo DE, Cameron DE, Naftel DC, Miller DC, Gillinov AM, Laschinger JC, Pyeritz RE (1999) Surgery for ascending aortic disease in Marfan patients: a multi-center study. N Engl J Med 340:1307–1313
21. Kim SY, Martin N, Hsia EC, Pyeritz RE, Albert DA (2005) Management of aortic disease in Marfan syndrome: a decision analysis. Arch Intern Med 165:749–755
22. Miller DC (2003) Valve-sparing aortic root replacement in patients with the Marfan syndrome. J Thorac Cardiovasc Surg 125:773–778
23. David TE, Armstrong S, Maganti M, Colman J, Bradley TJ (2009) Long-term results of aortic valve-sparing operations in patients with Marfan syndrome. J Thorac Cardiovasc Surg 138(4):859–864, discussion 863–4
24. Volguina IV, Miller DC, LeMaire SA, Palmero LC, Wang XL, Connolly HM, Sundt TM 3rd, Bavaria JE, Dietz HC, Milewicz DM, Coselli JS, Aortic Valve Operative Outcomes in Marfan Patients study group (2009) Valve-sparing and valve-replacing techniques for aortic root replacement in patients with Marfan syndrome: analysis of early outcome. J Thorac Cardiovasc Surg 137:1124–1132
25. Pyeritz RE (1989) Conference report: first international symposium on the Marfan syndrome. Am J Med Genet 32:233–238
26. Pyeritz RE, Loeys BL (2011) The 8th international symposium on the Marfan syndrome and related conditions. Am J Med Genet Dec 2 [Epub ahead of print]
27. Carson NA, Cusworth DC, Dent CE, Field CM, Neill DW, Westall RG (1963) Homocystinuria. A new inborn error of metabolism associated with mental deficiency. Arch Dis Child 38:425–436
28. Schimke RE, McKusick VA, Huang T, Pollack A (1965) Homocystinuria. Studies of 20 families with 38 affected members. JAMA 193:711–719

29. Hecht F, Beals RK (1972) "New" syndrome of congenital contractural arachnodactyly originally described by Marfan in 1896. Pediatrics 49:574–579
30. Beighton P, de Paepe A, Danks D, Finidori G, Gedde-Dahl T, Goodman R, Hall JG, Hollister DW, Horton W, McKusick VA, Opitz JM, Pope FM, Pyeritz RE, Rimoin DL, Sillence D, Spranger JW, Thompson E, Tsipouras P, Viljoen D, Winship I, Young I (1988) International nosology of heritable disorders of connective tissue, Berlin, 1986. Am J Med Genet 29:581–594
31. Beighton P, de Paepe A, Hall JG, Hollister DW, Pope FM, Pyeritz RE, Steinmann B, Tsipouras P (1992) Molecular nosology of heritable disorders of connective tissue. Am J Med Genet 42:431–448
32. DePaepe A, Deitz HC, Devereux RB, Hennekem R, Pyeritz RE (1996) Revised diagnostic criteria for the Marfan syndrome. Am J Med Genet 62:417–426
33. Loeys BL, Dietz HC, Braverman AC, Callewaert BL, De Backer J, Devereux RB, Holhorst-Hofstee Y, Jondeau G, Faivre L, Milewicz DM, Pyeritz RE, Sponseller PD, Wordsworth P, De Paepe AM (2010) The revised Ghent nosology for the Marfan syndrome. J Med Genet 47:476–485
34. Glesby MJ, Pyeritz RE (1989) Association of mitral valve prolapse and systemic abnormalities of connective tissue: a phenotypic continuum. JAMA 262:523–528
35. Dietz HC, Cutting GR, Pyeritz RE, Maslen CL, Sakai LY, Corson GM, Puffenberger EG, Hamosh A, Nanthakumar EJ, Curristin SM, Stetten G, Meyers DA, Francomano CA (1991) Marfan syndrome caused by a recurrent de novo missense mutation in the fibrillin gene. Nature 352:337–339
36. Loeys BL, Schwarze U, Holm T, Callewaert BL, Thomas GH, Pannu H, De Backer JF, Oswald GL, Symoens S, Manouvrier S, Roberts A, Faravelli F, Greco MA, Pyeritz RE, Milewicz DM, Coucke PJ, Camreon DE, Braverman AC, Byers PH, De Paepe AM, Dietz HC (2006) Aneurysm syndromes caused by mutations in TGF-β receptor. N Engl J Med 355:788–798
37. Williams JA, Loeys BL, Nwakanma LU, Dietz HC, Spevak PJ, Patel ND, Francois K, DeBacker J, Gott VL, Vricella LA, Cameron DE (2007) Early surgical experience with Loeys-Dietz: a new syndrome of aggressive thoracic aortic aneurysm disease. Ann Thorac Surg 83:S757–S763
38. Guo D-C, Papke CL, Tran-Fadulu V, Regalado ES, Avidan N, Johnson RJ, Kim DH, Pannu H, Willing MC, Sparks E, Pyeritz RE, Singh MN, Dalman RL, Grotta JC, Marian AJ, Boerwinkle EA, Frazier LQ, LeMaire SA, Coselli JS, Estrera AL, Safi HJ, Veeraraghavan S, Muzny DM, Wheeler DA, Willerson JT, Yu RK, Shete SS, Scherer SE, Raman CS, Buja LM, Milewicz DM (2009) Mutations in smooth muscle alpha-actin (ACTA2) cause coronary artery disease, stroke, and Moyamoya disease, along with thoracic aortic disease. Am J Hum Genet 84:617–627
39. Milewicz DM, Kwartler CS, Papke CL, Regalado ES, Cao J, Reid AJ (2010) Genetic variants promoting smooth muscle cell proliferation can result in diffuse and diverse vascular diseases: evidence for a hyperplastic vasculomyopathy. Genet Med 12:196–203
40. Pyeritz RE (2012) Evaluation of the adolescent with tall stature. Genet Med 14:171–177
41. Pyeritz RE (2013) Marfan syndrome and related disorders. In: Rimoin DL, Pyeritz RE, Korf BR (eds) Principles and practice of medical genetics, 6th edn. Elsevier, Philadelphia, (in press)
42. Bernhardt BA, Tumpson JE, Pyeritz RE (1992) The economics of clinical genetics services. IV. The economic impact of clinical genetics patients to the academic hospital. Am J Hum Genet 50:84–91
43. Giarelli E, Bernhardt BA, Mack R, Pyeritz RE (2008) Adolescents' transition to self-management of a chronic genetic disorder. Qual Health Res 18:441–457
44. Zayac C, Pyeritz RE, Bernhardt BA (2011) Why is genetic screening for autosomal dominant disorders underutilized in families? The case of hereditary hemorrhagic telangiectasia (HHT). Genet Med 13:812–820
45. Pyeritz RE (1981) Maternal and fetal complications of pregnancy in the Marfan syndrome. Am J Med 71:784–790
46. Rossiter JP, Morales AJ, Repke JT, Murphy EA, Pyeritz RE (1995) A prospective longitudinal evaluation of pregnancy in the Marfan syndrome. Am J Obstet Gynecol 173:1599–1606
47. Pyeritz RE, Fishman EK, Bernhardt BA, Siegelman SS (1988) Dural ectasia is a common feature of the Marfan syndrome. Am J Hum Genet 43:726–732

48. Foran JRH, Pyeritz RE, Dietz HC, Sponseller PD (2005) Characterization of the symptoms associated with dural ectasia in the Marfan patient. Am J Med Genet 134A:58–65
49. Ravitch MM (1949) The operative treatment of pectus excavatum. Ann Surg 129:429–444
50. Scherer LR, Arn PH, Dressel D, Pyeritz RE, Haller JA Jr (1988) Surgical management of children and young adults with the Marfan syndrome and pectus excavatum. J Pediatr Surg 23:1169–1172
51. Arn PH, Scherer LR, Haller JA Jr, Pyeritz RE (1989) Clinical outcome of pectus excavatum in the Marfan syndrome and in the general population. J Pediatr 115:954–958
52. Nuss D, Kelly RE Jr, Croitoru DP, Katz ME (1998) A 10-year review of a minimally invasive technique for the correction of pectus excavatum. J Pediatr Surg 33:545–552
53. Streeten EA, Murphy EA, Pyeritz RE (1987) Pulmonary function in the Marfan syndrome. Chest 91:408–412
54. Stokes DC, Wohl MEB, Wise RA, Pyeritz RE, Fairclough D (1990) The lungs and airways in achondroplasia: do little people have little lungs? Chest 98:145–152
55. Chow K, Pyeritz RE, Litt HI (2007) Abdominal visceral findings in patients with Marfan syndrome. Genet Med 9:208–212
56. Pyeritz RE (2008) Victor A. McKusick (obituary). J Clin Invest 118:3242–3243

Chapter 9
Homage for Victor McKusick

Nicholas Avrion Mitchison

It is a privilege and a pleasure to salute Dr. Victor McKusick as the creator of OMIM, a database second in value only to PUBMED itself as a tool for the scientific and medical community. This short contribution focuses on a few instances where OMIM has been (or could be) of special value.

First then OMIM has a special role as an extender of intellectual life. For a scientist, the first endpoint is usually loss of capacity at the bench: "where has that bottle gone that I just put down?" This seems to occur typically in the early 1970s, although there is of course a great deal of variation. From then on into the mid-1980s, a period extends in which work can continue at the desk but not at the bench. There is still plenty to do, and it is at this stage that there is leisure to write one's memoirs and with luck find among young people some who need help. It is also a stage at which there is time to make the most of the great data bases.

In my own case, I have used this period to review via PUBMED the need for dual targeting in any future immunotherapy of cancer. The point is that cancer cells have proved wonderfully adept at evasion, whatever the insults thrown at them. In this situation, the obvious need is for dual targeting. Rolf Kiessling calls this, "It takes two to tango". I like to call this need "$10^{-5} \times 10^{-5} = 10^{-10}$", to indicate the sort of variant frequencies that might be involved. This study needed 94 PUBMED references.

Another review concerned the frequency of the modes of inheritance of inherited diseases [1]. It needed 32 PUBMED references in the main text and 110 OMIM references in the appendix. There was no way that this study could have been possible without the use of OMIM. Let me dwell on how it was carried out. Our aim was to compare the frequency of the three modes of inheritance XL, AR and AD in the human population. We wanted to compare phenotypes that were so far as

N.A. Mitchison (✉)
University College London, 13 Framfield Road, London M5 1UU, UK

UCL Research Department of Immunology, Cruciform Building,
90 Gower Street, London WC1E 6BT, UK
e-mail: n.mitchison@ucl.ac.uk

K.R. Dronamraju and C.A. Francomano (eds.), *Victor McKusick and the History of Medical Genetics*, DOI 10.1007/978-1-4614-1677-7_9,

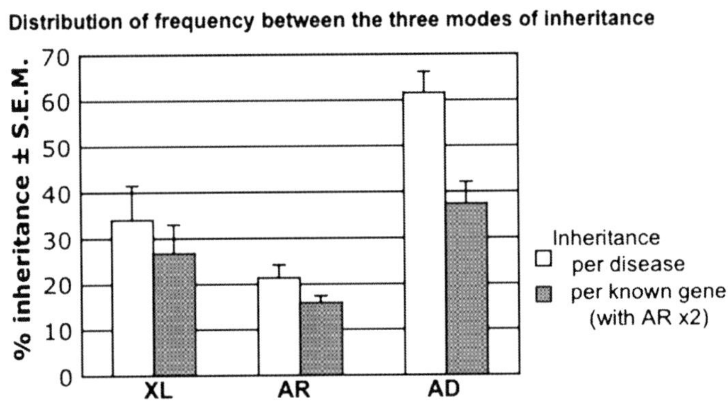

Fig. 9.1 Distribution of frequency between the three modes of inheritance

possible identical, in order to have the three types of disease subject to the same pressure of natural selection. In other words, the three disease phenotypes should be equally severe. We were worried about ascertainment bias and so needed a large number of comparisons obtained in the same way. The only way we could do so was to compare frequencies within the same disease, where the three phenotypes should be at least similar to one another. After all, diseases are discovered first, and the modes of inheritance only later.

To collect a sufficient number diseases with pairs (AR+AD) or triplets (XL+AR+AD) of disease genes, we carried out a search of OMIM using the limits and search facilities. This yielded 692 diseases, around a quarter of all diseases with molecularly defined disease genes. After further pruning, we were left with 95 diseases for the final statistical analysis. The outcome obtained in this way is shown in Fig. 9.1. It is immediately evident that there is a shortage of recessive inheritance, most notably in the percent recessive inheritance per gene. This is largely due to diseases where a relatively large (and often still growing) number of recessive genes have been discovered. All the four comparisons are statistically significant (how fortunate that the three-sigma level for significance demanded of the Higgs-boson-hunters is not required!).

So the question is: why is there so little recessive inheritance? We believe that two very different explanations are worth considering. One is simply that recessive diseases are more severe, so that those afflicted with the disease are less likely to reproduce. The second possibility follows a demographic scenario: up to the end of the last ice age, the human population was small and expanded as the climate warmed and agriculture began. This scenario has attractive features. Loss of recessive variation through inbreeding in small populations is well known. I believe that quantitative thinking about this process can be traced back to work in the Animal Breeding Research Organization (ABRO) led by Conrad Waddington from 1945. ABRO itself was modelled on the thinking behind Coastal Command, where several of the staff of ABRO had designed the tactics that saved the UK lifeline from the USA in the war that had just finished.

We calculated how long it would take for recessive inheritance to return, using a spreadsheet model to calculate the loss of recessives that continued during the period of recovery. The model of incomplete recovery, we believe, offers a reasonable explanation of the observed shortage of recessive inheritance.

At this point, I contacted an old friend, Robert Erickson of the University of Arizona at Tucson, hoping that the demographic scenario would appeal to him from his standpoint as an expert in the genetics of a tribal population. Not a bit of it. He suggested that disease severity might be the key: recessive disease might tend to be more severe.

So how to test this idea? It's far from clear to me how to search the databases systematically for information about the level of severity related to the mode of inheritance. This left me stumbling about, and I am still searching. Fortunately, an elegant set of data turned up, right in Dr. McKusick's home base at Johns Hopkins, where J A Winkelstein has long been running a registry of chronic granulomatous disease. The registry records that patients with the X-linked recessive form of the disease appear to have a more serious clinical phenotype than patients with the autosomal recessive forms of the disease, based on the fact that they are diagnosed significantly earlier (mean, 3.01 years of age versus 7.81 years of age), and have a significantly higher prevalence of a range of infections and other problems [2]. One swallow does not make a summer. Nevertheless I am pleased that the demographic scenario survives this first test and that this evidence argues against its alternative of severity.

A good place to look for further data on severity associated with the mode of inheritance might I hoped be the Orphanet database, which had already come in useful when testing the demographic scenario by providing a comprehensive list of rare genetic diseases ordered according to prevalence. Orphanet indeed aims to list diseases with their genes, beautifully crafted to show easily the mode of inheritance, frequency and severity in terms of age of onset. At least that is what they hope to do. Unfortunately, the information is at present too scrappy to be much use for my purpose. Some diseases, such as Charcot-Marie-Tooth disease, have useful sites that do indeed provide much of the relevant data (could this be because the discovery of CMT is one of Les gloires de la France?). But other sites provide only short texts that more or less just duplicate OMIM.

Let me briefly mention an interest in epistasis, a topic that OMIM will need increasingly to cope with. I have recently been concerned with epistasis, via an elegant study involving immunological tolerance induced in mice by de-aggregated bovine immunoglobulin [3], an old subject of mine. It turns out that susceptibility to this induction is under genetic control and that the gene responsible also influences susceptibility to autoimmune disease via an epistatic interaction [3]. A commentary written by Dr. A.M, Rose and myself makes the point that epistasis is fast becoming central to the understanding of the complex relationship between genotype and phenotype observed in autoimmune disease [4]. The question now is whether the concept of monofactorial disease will undergo widespread erosion and how OMIM will cope. So far, OMIM has a mere 60 references to epistasis, but this will surely much increase. Dr. Rose, I understand, plans to survey this issue.

There is a problem here. When raising the issue of epistasis with colleagues, they nearly always assume that this is a mistake and that I really mean epigenetics. That will have change.

In closing, may I recommend a wonderful film with a great actress that deals brilliantly, it seems to me, with a moral problem posed by Mendelian inheritance in man. I take the following account from the internet.

I've Loved You So Long (French: Il y a longtemps que je t'aime) is a 2008 French-language drama film written and directed by Philippe Claudel. It tells the story of a woman struggling to interact with her family and find her place in society after spending 15 years in prison.

When Juliette Fontaine (Kristin A. Scott Thomas) is released from prison after serving a 15-year sentence, her younger sister Léa invites her to stay with her family—including her husband Luc, his mute father Papy Paul, and their two adopted Vietnamese daughters, P'tit Lys and Emelia—in their home in the university town of Nancy in Lorraine. Léa, a college professor of literature, is considerably younger than Juliette. The younger woman recalls little about her childhood. Because of the nature of Juliette's crime (a secret which is revealed at the end of the film), their parents denied Juliette's existence and refused to allow Léa to visit her. In addition, Juliette had refused to speak throughout her trial. As a result, Léa knows nothing about the circumstances surrounding the crime, and when pressed for details, Juliette refuses to discuss what happened.

Gradually, Juliette begins to fit in with her family, makes friends and is given a permanent job. Léa then accidentally discovers a clue to Juliette's secret, leading to the film's final revelations: Juliette was a well-respected doctor; her young son had been diagnosed with a fatal and painful disease. When the disease progressed, Juliette had euthanized him so he wouldn't suffer. During her trial, she had felt so guilty for what she had done, she hadn't defended herself. Lea confronts Juliette with what she has learned and asks why she had never explained or asked for help, leading to an emotional breakdown between the sisters. Juliette, finally able to express her feelings and describe in detail what she did and why, is able to come to terms with the past and move on.

My only comment on the film is the fact the euthanized child is male is worrying. Surely, Juliette should have informed Lea? True that by a stroke of good fortune, Lea then had no children of her own, but how was Juliette to know that?

References

1. Mitchison NA, Bhattacharya S, Tuddenham EGD (2011) Human congenital diseases with mixed modes of inheritance have a shortage of recessive disease. A demographic scenario? Ann Hum Genet 75:688–693
2. Winkelstein JA, Marino MC, Johnston RB Jr, Boyle J, Curnutte J, Gallin JI, Malech HL, Holland SM, Ochs H, Quie P, Buckley RH, Foster CB, Chanock SJ, Dickler H (2000) Chronic granulomatous disease. Report on a national registry of 368 patients. Medicine (Baltimore) 79:155–169

3. Fujii T, Hou R, Sato-Hayashizaki A, Obata M, Ohtsuji M, Ikeda K, Mitsui K, Kodera Y, Shirai T, Hirose S, Nishimura H (2011) Susceptibility loci for the defective foreign protein-induced tolerance in New Zealand Black mice: implication of epistatic effects of Fcgr2b and Slam family genes. Eur J Immunol 41:2333–2340
4. Mitchison NA, Rose AM (2011) Epistasis: the key to understanding immunological disease? Eur J Immunol 41:2152–2154

Chapter 10
How Did My Association with Victor Influence Me, My Career, and My Family?

Digamber S. Borgaonkar

Our second child, Sonya, was born in Pune (India), in 1966. The obstetrician suspected something was abnormal, and on a pediatrician's advice, X-ray of the lower extremity was sent to me in Baltimore. Congenital bowing of the femurs was a likely diagnosis, but Victor suggested a whole-body X-ray be done. Accordingly, this was done and read as "within normal limits." Few weeks passed by before she fractured her right hip with a slight trauma. Only then the entire set of X-ray films were sent to me. Victor took one look at the skull X-ray film and said "the radiologist does not know what he is talking about." Immediately, Victor gave me an autographed copy of his "Heritable Disorders of Connective Tissue" with a notation "With Highest Regards" and told me that the film of Sonya's skull X-ray is better than the ones in this book! I was stunned. In those days, we, medical geneticists, did not have clear understanding as to the modes of inheritances of different types of OI and their characteristics. Manda and I did belong to the same "Gotra" in a Maharashtrian Brahmin family [1]. Whether Sonya had a severe form of recessive OI or a mild form dominant OI was a big question for us. Manda had not gone to her parent's home in India for the birth of our first child (Raj) as is the custom in Indian society mainly because I wanted my child to be an American citizen by birth! Sonya came along 15 months later, and for convenience, she went to her father's house in Pune. Victor visited that house in 1971 and talked about it fondly since it was 300 years old! Around the same time, my work on the chromosomes of Tenrecs of Madagascar, funded by the National Science Foundation, was becoming very interesting. I was likely to go there to collect some animals [2]. Somehow, Manda, agreed to stay in India for few more months, and we flew back to Baltimore together in

D.S. Borgaonkar (✉)
Laboratory of Neurogenetics, National Institute of Aging, National Institutes of Health, Bethesda, MD 20892, USA

11 Stabler Circle, Wilmington, DE 19807, USA
e-mail: dsborgaonkar@hotmail.com

K.R. Dronamraju and C.A. Francomano (eds.), *Victor McKusick and the History of Medical Genetics*, DOI 10.1007/978-1-4614-1677-7_10,

1967 with both of our children. Victor helped in many decisions that we had to make in raising Sonya, her many surgeries, schooling, and general upbringing. Sonya now is a practicing attorney in Virginia and has a typical suburban lifestyle with a dog, house with two children, and a commuting (to DC) husband. We were very fortunate in being able to send her to the Bryn Mawr Lower School which had just opened a brand new building for their lower school. This one had no steps what so ever. During the first year, Sonya was not able to walk, nor crawl, but used to scoot on her buttocks. By then, she had had two osteotomies. Dr. Robert Abrams did those procedures at the Kernan Hospital in Baltimore, where sports luminaries like Johnny Unitas had theirs there, even though on staff at JHH. But by the time Sonya started school, all the foreign bodies had to come out of her body! Her radiologist, John Dorst, once called me back after saying Hi and enquired about Sonya. He had remarked that her X-ray of the hip looks like an X-ray of a "cardboard." A physician relative, from India, remarked to me few years ago that upbringing of Sonya in India would have been different, meaning difficult if not impossible. Needless to say that all this would not have been difficult even in this country had I not been at Hopkins in Victor's department. I owe him a lot. While Sonya was working one summer at the NIH's ELSI program, she attended a meeting. Other attendees were surprised to see Victor and Sonya were engaged in a deep conversation wondering what they were talking about!

During the late 1960s and after the popularization of the phenotype of 47,XYY males, I had a very good fortune of having "picked up" quite a few such individuals in cases referred to us for diagnostic workup. I felt strongly that we should undertake a population survey of some sort to determine the proportion of such individuals in different populations. I prepared a grant application for support with various colleagues from JHU to survey institutions for delinquent boys, county schools, and pediatric clinic populations. Victor was supportive and would accompany me to meetings with Baltimore County school officials, meetings with pediatric department officials including Bob Cooke, Henry Seidel, Shirley Borkowf, and others. He also pointed the need for him to be there at the meeting with Maryland State Medical Society since I was not a member (not an M.D.). We "sailed" through all such formalities because of VAM's support [3]. Within the NIH system, however, there was some discussion as to where this work would best "fit." The NIMH was already involved in supporting studies on 47,XYY (Stan Walzer at Harvard). Dr. S. Shah, a clinical social psychologist, made a case and supported our application for funding in the fall of 1969. Soon after the study had started, there was some bad publicity, because of erroneous misconceptions about the study which were published in local papers, resulting in unexpected problems. Fortunately, those problems were overcome and the study progressed. Victor was always there to support me and had fed accurate information to other higher up administrative authorities at Hopkins. When I was being introduced to give a talk at a Department of Medicine meeting, the revered chairman, A McGehee Harvey, with whom I never had one-to-one meeting, knew all the details about the study, the problems it had encountered! Source had to be Victor [4].

The work on computerization of chromosome abnormalities began in 1974. Knowing Victor's interest in mapping, I proposed that I go to Frank Ruddle's lab with the cell line in which chromosome 14 was split in two in the middle because of a translocation between chromosomes 5 and 14. I had picked up this family in a study [5]. The gene NP for nucleoside phosphorylase had been assigned to chromosome 14 earlier. I commuted to New Haven regularly in 1974. While at Yale, I conceived the idea to computerize all of the known chromosome abnormalities in a database. Incidentally, an interesting historical fact: The establishment of Yale College had something with India's richness, Eli Yale, was a governor of Madras presidency in the colonial British era. I remember describing to Victor, one night about my idea to compile the database. He heard me and commented, "start writing the book!" By the year end, I submitted the manuscript of the first edition to the Johns Hopkins University Press [6]. Even though I had to part ways with Johns Hopkins, a second edition was published by Alan R. Liss in 1978, and in subsequent years, several editions were published till 1997. The online version of "Chromosomal Variation in Man" has been available since 1998 at www.wiley.com/borgaonkar.

I do not think this would have been possible for me to accomplish had I not been at Hopkins. With the exception of one review, all others, more than several dozen, of Chromosomal Variation in Man have been nothing but complimentary!

There were two other databases that I got involved with. One was The International Registry (later changed to "Repository") of Chromosomal Abnormalities in Man. Several Listings of this book were published by me from the institutions where I worked. This project got me into some controversial settings (One of the book reviewers of the first edition of "Chromosomal Variation in Man," wrote an excellent review but stated in her letter to me, in 1975, "So, it was not until the recent ASHG meetings that I became aware of existing controversies with other registries and ... "). But, I continued it under a different name, Repository of Human Chromosomal Variants and Anomalies till 1993 when I published the 14th Listing. The International Registry project received funding from the World Health Organization. It also enabled me to visit many countries under the sponsorship of the Fogarty International Center of the NIH including the Soviet Union in 1978 and many other countries under the National Academy of Sciences' Scientist Exchange Programs. The Chinese invited us in 1985, and they published their registry later. Based on my reports to the NIH and the NAS of the visit to the USSR, I was told that it was at the top of the reports submitted by exchange scholars while evaluating whom to invite for a meeting to be arranged in Washington, DC. This was being done at the urging of President Reagan before his meeting with the Soviets to reopen the dialogue after he had opined them as evil empire, etc. I was flat on my back in June 1984, after having had a heart attack, when the organizers had called me! Because of my condition, I attended this meeting at the Stone House, Fogarty International Center, NIH, Bethesda. I had predicted in my report that the Soviet system is superficial, the country is not that strong and their facilities are primitive, although we had a good time over there [6, 7, 8].

Later, a cancer chromosome abnormality registry was planned with support from the International Union. A registry was planned after arranging a meeting at Hopkins in 1977 with substantial support from John K Frost, a cytologist in the Pathology department with whom I had started to collaborate since 1970 because his lab had a fluorescent microscope and the then new banding technique required its use! Many if not all the then leading cancer cytogeneticists attended this meeting. I remember Victor giving them a tour of the main hospital building and taking them around the famous "Dome." After my departure from Hopkins in 1978, I did not see any value for me to continue this cancer registry.

There is no question in my mind that my years in Baltimore in Victor's group changed my professional life for good and have enjoyed the benefit in working in human genetics field for almost half a century. Taking the course of human genetics of Sheldon Reed at the University of Minnesota in my first year of graduate school in USA in 1959 gave me the idea to write to Victor about the possibility of working with him. Initially, he did not need me but when Peter Bowen left Hopkins to start the Medical Genetics program in Alberta, my second enquiry had reached his desk and the rest is history!

In the photograph: front row (seated): Second person is Ms. Margaret Abbott, 3rd is Dr. Tony Murphy, 4th is VAM, 5th is DSB and 6th person is Dr. Judy Hall

References

1. Borgaonkar DS (1990) Prevention of consanguineous marriages in ancient India. Bionature, pp 1–4
2. Borgaonkar DS (1969) Insectivore Cytogenetics In Comparative Mammalian Cytogenetics, Ed. K. Benirschke, Springer-Verlag, NY, pp 218–246
3. Borgaonkar DS, Shah SA (1974) The 47,XYY chromosome male- or syndrome? Prog. Med. Genet. 10: 135–222
4. Borgaonkar DS (1978) Cytogenetic screening of community-dwelling males. In Genetic issues in Public Health and Medicine, Eds. Cohen BH, Lilienfeld AM, Hunag PC, CC Thomas, Springfiled, pp 215–234
5. Borgaonkar DS, Blair SM, Lutz JB, Kelly TE, Tice RR, Delaney N, Hurchinson JR, Bias WB (1973) Cytogenetic study of a 5/14 translocation in man. J. Hered. 64: 299–300
6. Borgaonkar DS, Bolling DR, Blair S, Classon R (1975) International registry of abnormal karyotypes. Provisional listing. The Johns Hopkins University School of Medicine, Baltimore, 85 pp
7. Borgaonkar DS, Lillard DR, Kamat EB, Khaldi AS, Putnam C, Aiken K, Clark M, White O, Ziegenfus T (1980) Repository of chromosomal variatiants and anomalies in man- an international registry of abnormal karyotypes. 7th listing. Genetics Center/North Texas State University, Denton
8. Borgaonkar DS, Reisor N, Ahn J, Caldwell R, Gahunia H, Pasquenza N, Petersen E, Wellmon BD (1993) Repository of human chromosomal variants and anomalies-an international registry of abnormal karyotypes. 14th listing. Medical Center of DE and University of Delaware, Newark, 879 pp

Chapter 11
The Assimilation of Classical Genetics into Human Genetics

Elof Axel Carlson

The Separate Origins of Classical, Human, and Medical Genetics

Classical genetics may be considered as that body of knowledge of heredity that developed between 1900 and 1920. It includes the rediscovery of Mendelian laws of transmission, the chromosome theory of heredity, the discovery of X-linked traits, the discovery of recombination by crossing-over during meiosis, the discovery of chromosome nondisjunction, the mapping of the known genes in fruit flies, maize, and other laboratory organisms, and a variety of chromosomal rearrangements, both gross and minute. It also includes gene-character relations, the inheritance of quantitative traits, and the demonstration of multiple factor inheritance. Underlying these findings, which emanated largely from the work of E. B. Wilson and his students, T.H. Morgan and his students, and E. A. Emerson and his students in the United States, were the work of William Bateson in England, Hugo DeVries in Holland, Carl Correns in Germany, Wilhelm Johansson in Denmark, and Herman Nilsson-Ehle in Sweden. Before the rediscovery of Mendelism, what was later called genetics, was largely a mathematical activity carried out by Darwinists in Great Britain, especially those influenced by Francis Galton, Karl Pearson, and W. F. R. Weldon. They called their approach to heredity, biometrics. Opposed to them was the work of William Bateson in the 1890s who saw evolution requiring discontinuous inheritance that he called homeotic and meristic. Ironically both the biometric school and Bateson's school ignored the chromosome-based classical genetics that was emerging, especially in the United States [1].

E.A. Carlson (✉)
Stony Brook University, Stony Brook, NY, USA

Institute for Advanced Study, Indiana University, PO Box 8638, Bloomington, IN 47407, USA
e-mail: ecarlson31@netzero.com

K.R. Dronamraju and C.A. Francomano (eds.), *Victor McKusick and the History of Medical Genetics*, DOI 10.1007/978-1-4614-1677-7_11,

Human and medical genetics did not exist as such during this period, from 1890 to 1920. Human heredity was embedded in degeneracy theory from roughly 1700 to 1883. In 1883, Galton coined the term eugenics to represent the study of the science of human heredity and its improvement or preservation. It was wrapped in social values, unlike classical genetics, which both in its pure and applied forms was largely uninterested in social issues. Economic issues emerged later with the development of hybrid corn and other commercial applications of genetics, but those took place mostly after 1930. Prior to these "agribusiness" approaches, applied classical genetics was the realm of hobbyists and individual breeders. Ironically, the most successful large-scale applications of breeding for commercial gain before the 1930s were carried out by I. V. Michurin in the USSR and Luther Burbank in the United States. Both used selection, not Mendelian genetics, and both considered themselves Lamarckist not Mendelian in their scientific outlook [2].

The eugenics movement served as a large umbrella covering social reformers, socialists, utopians, physicians, racists, ethnic bigots, class elitists, and a variety of intellectuals. They also embraced geneticists, whether they were Mendelian or Lamarckist. Most shared a "like for like" model of inheritance in which healthy individuals bred healthy children and degenerate or unfit individuals bred degenerate or unfit children. By the beginning of the twentieth century, the eugenics movement split into two wings. One group, led by Galton, favored "positive eugenics" in which those with higher intelligence, talents, good health, longevity, and caring personalities were asked to have more children than the average person of their status. In Galton's mind, this would lead to an improvement of humanity incrementally each generation and greatly stimulate progress and prosperity for everyone as well as stimulate civilization to greatness. The other group, largely developed in the United States and led by Charles Davenport at Cold Spring Harbor in New York, favored "negative eugenics" in which the American stock was threatened by those with unfit protoplasm, who would debase the population through their larger family size and their inherited social failures leading to pauperism, insanity, feeblemindedness, and immoral behavior. In 1905, a third split occurred when Alfred Ploetz in Germany coined the term race hygiene to represent a negative eugenics movement directed specifically at cleansing impurities from an alleged ideal of the ethnic group to which those geneticists belonged. Its stress was producing healthy children in an assumed superior or desired race. By the 1930s, it dominated German eugenics and was taken over by the Nazi Party [3].

This historical overview of classical genetics in relation to human and medical genetics before 1920 provides the broad worldview of heredity at that time. If we contrast that older view with today's status of classical genetics, human genetics, medical genetics, and molecular genetics, we would find a very different set of assumptions [4]. In today's thinking, geneticists are comfortable reading the literature and seeing the connections of classical, human, medical, and molecular genetics. They differ in their tools. Classical genetics is still guided primarily by breeding analysis and genotype-phenotype relations. Human genetics seeks the basic genetics of humans but relies on pedigree analysis and mathematical models of gene relations for mapping and for gene distribution in populations. Medical genetics relies on the tools of diagnostic medicine for identifying specific syndromes or conditions from cognate disorders. It also has as its mission such applications as genetic screening, prenatal diagnosis, and therapy where possible. Molecular genetics applies tools of DNA analysis and nucleotide sequencing to identify genes and their mutations applied to classical genetics, human genetics, or medical genetics. While most of these changes occurred after 1953 with the identification of the structure of DNA, there were efforts to bridge these disciplines throughout the twentieth century. It is worthwhile to look at some of the attempts to bridge classical and human or medical genetics in that pre-DNA era to show how the thinking shifted from approaches used in classical genetics to those needed for human or medical genetics to assimilate the new findings.

X-Linked Inheritance and Gene Mapping

The recognition of sex chromosomes began with the work of H. Henking in 1891 who described an unusual chromosome-like object in the fire-wasp that differed in size, location, and movement during cell division. He called it an X element, using X for the mathematician's symbol for an unknown number. Some 10 years later, Clarence McClung associated Henking's finding with one of his own in grasshoppers that he called the accessory chromosome and which appeared only in males. Wilson and independently Nettie Stevens in 1905 reported a more complex relation that she called heterochromosomes and he called idiochromosomes associated with the sex determination in these insects. Wilson renamed them as X and Y chromosomes and noted that most of his species were XX female and XO or XY males [5]. In 1910, Morgan found a white-eyed mutation that provided an altered Mendelian ratio of three red-eyed flies (with a two female to one male ratio) to one white-eyed male. His finding of two more sex-associated mutations, miniature wings, and rudimentary wings convinced him that all three of these mutations were associated with the X chromosome. Wilson then suggested to Morgan that in humans, red-green color deficiency must also be an X-linked trait. Morgan followed up his X-linked traits and showed they gave unusual ratios that he interpreted as crossing-over between these genes in the pairs of X chromosomes carrying them. His student Alfred Sturtevant, in 1913, used Morgan's data to construct a map of five X-linked factors: yellow body color, white eyes, eosin eyes (an allele of the white factor that arose in a stock bottle of white-eyed flies), miniature wings, and rudimentary wings.

While Davenport and others recognized hemophilia (later, factor 8 variety) as also being X-linked in humans, mapping of factors on the human X did not take place until JBS Haldane and Julia Bell in 1937 used pedigrees looking for recombination between these two factors. They reported evidence of crossing-over from pedigrees of several patients showing the combination or separation of these factors. Haldane and Smith in 1947 estimated a ten map unit separation, but later studies of the most common hemophilias, type A and B, showed a much lower frequency of recombination, that between gene F8 (the hemophilia A form which is the most commonly encountered type) and the red-green locus [now designated as OPN1MW] virtually void of recombinant events. Both are in the Xq28 region (but separated by about 20 known genes) [6].

Mapping by pedigree analysis was difficult, and the number of useful families where two or more genetic factors could be followed was rare. For this reason, useful maps of human genetic traits did not emerge until cytogenetics was incorporated into human genetics. Unlike fruit flies with a diploid chromosome number of 8, humans were thought to have 46–48 chromosomes in the first half of the twentieth century. In 1956, that number shifted to 46 with the work of Tjio and Levan [7]. The discovery of nondisjunctional disorders (especially Down syndrome, 47,+21; Klinefelter syndrome 47,XXY; and Turner syndrome 45,X) by 1960 accelerated the shift of human genetics toward medical genetics. Even more important was the discovery of Q and G banding of chromosomes which allowed a cytological approach to mapping genes. The presence of deletions, duplications, inversions, and translocations made genes near the breakage points of these events mappable. Many techniques emerged to map chromosomes by hybridization of human cells with other mammalian cells. But the most dramatic change came with the technologies for sequencing DNA and using chromosome fragments to locate genes and order them into a sequence. By the 1990s, the possibility of a complete genome sequence for humans was theoretically possible, and by the start of the twenty-first century, it became a reality [8]. The molecularization of genetics simultaneously made breeding analysis for mapping genes in classical organisms and mathematical or cytogenetic approaches to mapping in humans obsolete. In a similar way, what was considered new in 1955, the introduction of genetic fine-structure mapping by Seymour Benzer in phage and Milislav Demerec in bacteria was also rendered obsolete. Fine-structure mapping used the conceptual tools of crossing-over for microbial mutations and revealed the parallel between that fine-structure and the assumed DNA nucleotide basis for the numerous sites within a gene (cistron). The direct sequencing approach of genes and chromosomes eliminated the need for a breeding analysis to uncover map sequences. It leveled the playing field for geneticists across all phyla.

Aneuploidy in Humans Needed New Techniques

Aneuploidy was first recognized by Ruggles Gates when he was studying Oenothera (the evening primrose), but he did not propose a theory of nondisjunction for its origin. He observed variations in chromosome numbers from strains of primroses he crossed. The idea of nondisjunction was promoted by Calvin Bridges in 1916 as he worked through the combined genetic expression of the white-eyed alleles in crosses associated with their nondisjunctional origin in fruit flies [9]. Thus, if a white-eyed male is mated to a red-eyed female, one expects all red-eyed offspring in the F1. But occasionally, Bridges encountered a white-eyed male, and it was always sterile. In the reciprocal cross, a red-eyed male mated to a white-eyed female should give "crisscross" inheritance with F1 showing white-eyed sons and red-eyed daughters. But occasionally, Bridges found a white-eyed daughter. These females turned out to have a Y chromosome that they could sometimes supply to their offspring. Bridges used nondisjunction as evidence for the chromosome theory of inheritance. Bridges had no difficulty showing the cytological outcomes and correlating them with their genetic expression.

Human cells were different because they had a large number of chromosomes crowded in a nucleus (about six times as many as fruit flies). T. S. Painter thought the number could be 46, 47, or 48 but settled on 48 along with a definitive X and Y. That was 1926 and did not change for almost 30 years when the number shifted permanently to 46 as our species number. The idea that there would be chromosomal expressions of disease in humans had a long history. Abnormal chromosome numbers were observed in aborted and abnormal sea urchin embryos that were induced by agitation by Theodor Boveri in 1904. Later, in 1914, he proposed a theory of tumor cell origin and cancer development associated with aneuploidy. By 1956, techniques had improved, and Tjio and Levan provided them for others to confirm their finding and to study possible candidate disorders for a chromosomal connection. Lejeune and Turpin in 1959 were the first to do so with the identification of Down syndrome as a trisomy 21. Very rapidly, additional conditions were found. I believe one reason a search was not made earlier was the assumption that analogous cases in humans to the XXY and XO in flies would be fertile females and sterile males, neither of which would stand out. Also, autosomal nondisjunction in fruit flies was incompatible with survival. The discovery that the Y is male determining in mammals including humans did not emerge until the 1960s, and it came as a surprise to fruit fly geneticists who generally regarded the Y as limited to male fertility genes [10]. This expectation of universality from genetic findings in flies or maize to humans made it both difficult to find in humans and a surprise when it emerged in a novel way.

Dosage Compensation Revealed Two Dissimilar Mechanisms Existed

Just as the sex chromosome nondisjunctions in flies and humans differed in consequences on sexual expression, so too did dosage compensation differ for flies and humans (it was found in flies independently in 1932 by Muller's laboratory then at Texas and by Curt Stern in Germany). Muller named the phenomenon and demonstrated that it required dosage compensating genes which he isolated using deleted X chromosomes that carried them. For the human mechanism of dosage compensation, the story evolved out of the discovery of sex chromatin by Murray Barr studying cells in cat brains. Sex chromatin was then extended to human cells. It was then found to be contradicted by sexual expression in cases of Klinefelter syndrome and Turner syndrome. It was then resolved by Mary Lyon who identified the sex chromatin as X-inactivation [11]. While this cleared up the contradictions of sex expression and dosage compensation in mammals, it differed from the mapping history because there is a very different mechanism in fruit flies and mammals. A molecularization of the two interpretations of dosage compensation into a common ground as in mapping by nucleotide sequence is not possible. They have independent evolutionary histories. In flies, compensasomes (dosage compensation complexes) are involved to boost the male output of the X chromosome to levels found in females with two X chromosomes. In mammals, the Xist gene is involved in inactivating one of the two X chromosomes in females to render each female cell to the same dose level as that of a male with a solitary X.

Gene Structure as It Is Assumed and as It Is Revealed

In my own career, I ran into this uncertain outcome of assimilation of classical genetics with human and medical genetics. As I wrote up my dissertation in 1958 on the structure of the dumpy locus in fruit flies, I studied carefully the literature on gene structure going back as far as I could. For fruit fly geneticists, the prevailing model was called pseudoallelism mostly worked out by E. B. Lewis. His findings used an evolutionary model that Muller and independently Bridges promoted in which genes underwent an initial tandem duplication, and this permitted each of the duplicated genes to mutate independently and acquire cognate functions. Opposed to this was the developing model of intragenic recombination favored by G. Pontecorvo using the fungus Aspergillus and by S. Benzer using bacteriophage. In my own work, I found both phenomena, the dumpy set of inferred duplications also showing intragenic recombination for one particular dumpy component (what I called dp-ov and what Bridges found and called dumpy). I then noted that a similar conflict of interpretation existed for the Rh blood groups in humans. A.S. Wiener, working with K. Landsteiner, had designated the gene as Rh and its recessive form as rh. There were numerous alleles of these found by tests of their immunological specificities with other Rh or rh alleles. The Wiener system was difficult to follow and required a lot of immersion in the terminology, just as my own early immersion in the dumpy alleles (designated as lopped, thoraxate, truncate, drumlins, vortices, comma, etc.). I simplified my system by using the three major phenotypes involved: o = oblique wings, v = thoracic vortices, and l-recessive lethal. This allowed me to combine the expressed traits of an allele as ov, olv, lv, ol, v, o, l for the older terminology of dumpy, truncate, thoraxate, lopped, vortex, oblique, and dumpy-lethal. It was much easier to follow than the unique names assigned to these alleles when they first arose. I did not know at the time I did this that R.R. Race and R.A. Fisher had developed a

CDE or cde system for the Rh series, but as I read those papers, I realized that the conflict between the cde and the Rh system of notation could be resolved if the Rh system was a tandem sequence of genes within which mutations could produce a plethora of allelic types. I wrote that up as an article for the *American Journal of Human Genetics,* although one referee was worried I would be drawn into a debate that might hamper my career. Years later, the sequences were worked out [12]. As is often case when geneticists compare "phenogenetic" approaches to molecular approaches, the molecular findings are more accurate and often reveal aspects not predictable by phenogenetics. There are two very similar tandem genes, RHD and RHCE located in 1p34–36.2 Each has ten exons. The RHD is flanked by "rhesus boxes" that virtually are homologues and which lead to occasional mispairings and the loss of the RHD gene (and hence, the Rh negative expression). In addition to the two proteins made by the RHD and the RHCE genes, there is a gene on chromosome region 6p11–21, RHAG, which produces a protein that combines with the RHD and RHCE proteins to produce a tetramer that inserts into the red blood cell membrane and allows either ammonia transport or carbon dioxide transport into and out of the cell. If there is a mutation or loss of the RHAG gene, no Rh antigen is formed in the red blood cells, and such individuals have a mild anemia and relatively fragile cells that can lyse. While my initial assumption about reconciling the debate by Weiner and Fisher-Race was correct, the details of the way the antigen forms and the basis for the origin of the Rh negative condition was not predictable from an analysis of the known allelic types or the disorders they produced.

Retinoblastoma and Genetic Counseling

My second attempt at imposing a classical genetic model on human and medical genetics involved retinoblastoma. I had developed a model of mutational mosaicism using fruit flies and the dumpy locus. When I induced such mutations with X-rays, the mutants that emerged were "complete" in the sense that when bred to a standard dumpy fly, they produced the allelic variant of dumpy seen at its origin. It was typically Mendelian with half the offspring showing the new mutant allele. But when I used quinacrine mustard as a mutagen to damage sperm, the emergent allele was often "fractional" or mosaic in first emergence. Thus one wing might be normal and the other wing might show severe truncation, and on that same side as the mutant wing, there might be bristle disturbances and a thoracic vortex. When I bred such mosaics, only about 16% transmitted the mutation to their offspring and those showed the new defect throughout their body and no longer showed mosaicism in subsequent generations. I interpreted this as an outcome of the distribution of mutant and nonmutant cell lines in the embryo formed by the altered sperm DNA. The chemical had altered a base on one strand of the double helix, and after fertilization by such a sperm, two cell lines emerged: one normal and one bearing the new dumpy allele. The fact that most fractional mutations did not transmit suggested that the mutant cells did not get incorporated in the germ line as often as they were present in part of the soma. My student, John Southin did an extensive analysis of these combinations of somatic and gonadal involvement.

A friend of ours had a son with retinoblastoma when I was a young professor at UCLA in 1961. He had survived a bilateral condition but lost both eyes and was now entering his teens. I looked up retinoblastoma in the biomedical library at UCLA and found many parallels to the dumpy story. I did not write it up, however, until many years later, when I spent a sabbatical at the University of Minnesota and discussed my interest in retinoblastoma with Robert Desnick. He gave me access to the medical files of their cases and I worked out a paper with Desnick on the genetic parallels with retinoblastoma and dumpy and I also worked with Desnick to prepare a paper on genetic counseling [13]. I learned from talking to some parents with children who had a child with unilateral retinoblastoma that they had been pessimistically counseled; they were told it was a classical dominant mutation, and half of the children would be affected. I knew this was false from the survey of patients' records at the University of Minnesota and from numerous papers reporting retinoblastoma around the world.

What made retinoblastoma so unusual was its transmission when familial and usually bilateral. It acted as a typical autosomal dominant mutation with a risk of 50% transmission for each pregnancy. Yet the most likely interpretation of the origin of the tumor cell leading to retinoblastoma in an at-risk infant or child was a second hit to an appropriate cell in a child who was essentially heterozygous [14]. In a sense, the dominance is a homozygous recessive cell caused by the second hit in a heterozygous retinal precursor cell in the fetal or infant eye.

Today retinoblastoma, RB1, has been mapped to 13q14.2, and it is a large gene with 27 exons. It produces a metal-binding protein that serves as a regulator of genes active in the S phase of the cell cycle. It normally represses cell division by binding to a transcription factor, but in the mutant condition, it will fail to do so, and this leads to tumor growth. But retinoblastoma also acts as a tumor-inducing gene in other tissues, especially the pineal gland (a modified eye in its evolutionary history), and osteosarcoma. These secondary tumors are more likely to be encountered in new bilateral cases or in familial cases. They would be less likely in unilateral cases if the unilateral child has fewer cells carrying the recessive RB1 gene mutation in its somatic tissues. Also complicating the counseling is the penetrance of the RB1 gene. For bilateral cases, it is high, about 95%, showing one or both eyes involved, mostly both eyes. For the unilateral sporadic cases, the counselor is dependent on empiric risk estimates (less than 20%) with a likelihood that it will express as a bilateral case if the new child develops the condition. What is not certain is whether tests of skin cells and blood would reveal the existence of the mutant gene in a young adult seeking counseling information for a successfully treated unilateral retinoblastoma. The absence of such somatic cells does not automatically exclude germinal involvement. The molecular sequence of the RB1 gene, however, does permit prenatal diagnosis which was not possible in the classical genetics era.

What Role Does Classical Genetics Play in the Future of Human and Medical Genetics?

There are many connections for classical genetics to enter the new molecularized human and medical genetics. In forensics DNA, analysis has been widely popularized for solving paternity and criminal cases and for identifying suspects involved in espionage or identifying the bodies of terrorists and other targeted figures in war. It has largely replaced fingerprinting, dental records, or blood groups where DNA is available (especially in tooth pulp and bone where it is more likely to survive bacterial decomposition). In evolutionary studies, the DNA of mitochondria and Y chromosomes has been supplemented by nuclear DNA from mummified or preserved tissue, especially hair. Anthropologists have used this to trace the migration of humans from Africa 100,000 or more years ago throughout the world. The sequencing of Neanderthal genes has also proven successful in showing admixtures of Neanderthal genes in subtypes of our species. The human genome project is still in its infancy, but eventually, McKusick's catalog of single-gene disorders will include the mapping and sequencing references for all of them, as well as their role in the cell, development, or physiology of the individual. In a reversal of the pattern of history, McKusick's OMIM [Online Mendelian Inheritance in Man] is a return to human genetics from molecular medical genetics. The *Treasury of Human Inheritance* that Karl Pearson started and Julia Bell largely filled with single-gene case histories and pedigrees was the progenitor of McKusick's enormously valuable *Mendelian Inheritance of Man* and its present status as OMIM [15].

Herman J. Muller 1927

References

1. For a history of classical genetics, see Carlson EA (2007) Mendel's legacy: the origin of classical genetics. Cold Spring Harbor laboratory Press, Cold Spring Harbor. For a history of Bateson's involvement in the dispute with biometricians, see Cock A, Forsdyke DR (2008) Treasure your exceptions: the science and life of William Bateson. Springer, New York
2. For histories of human and medical genetics, see Dronamraju K (1989) The foundations of human genetics. Charles C Thomas, Springfield. Harper PS (2008) A short history of medical genetics. Oxford University Press, New York
3. For a history of degeneracy theory and its relation to eugenics, see Carlson EA (2001) The unfit: a history of a bad idea. Cold Spring Harbor Laboratory Press, Cold Spring Harbor
4. For a history of molecular genetics see Olby R (1974) The path to the double helix: the discovery of DNA. University of Washington Press, Seattle. Judson HF (1979) The eighth day of creation: makers of the revolution in biology. Jonathan Cape, London (now Random House, NY)
5. For a detailed history of the discovery of sex chromosomes see Carlson op. cit., 2007 Chap 7 The sex chromosomes, pp 79–98
6. Bell J, Haldane JBS (1937) The linkage between the genes for colour-blindness and hemophilia in man. Proc Roy Soc B 123:119–150. Haldane JBS, Smith CAS (1947) A new estimate of the linkage between the genes for colour-blindness and hemophilia in man. Ann Eugenic 14: 10–31
7. Kottler MJ (1974) From 48 to 46: cytological technique, preconception, and the counting of human chromosomes. Bull Hist Med 48:467–471
8. Collins FS (2010) The language of life: DNA and the revolution in personalized medicine. Harper Collins, New York
9. Gates RR (1909) The stature of chromosomes of Oenothera gigas, De Vries. Archiv für Zellforschung 3:525–552. Bridges CB (1916) Nondisjunction as proof of the chromosome theory of heredity. Genetics 1:1–53; 107–163
10. For a pre-molecular history of the human X and Y chromosomes, see Mittwoch U (1967) The sex chromosomes. Academic Press, New York
11. Barr ML, Bertram EG (1949) A morphological distinction between neurons of the male and female and the behavior of the nucleolar satellite during accelerated nucleoprotein synthesis. Nature 163:676–677. Also, see the review by Lyon M. The William Allen White Memorial Award Address. X-chromosome inactivation and the location and expression of X-linked genes. Am J Hum Genet 42:8–16
12. Carlson EA (1958) The bearing of a complex-locus on the interpretation of the Rh series. Am J Human Genet 10:465–473. Race RR, Sanger (1954) Blood groups in man. Charles Thomas, Illinois. Wiener AS (1954) Rh-Hr blood types. Grune and Stratton, New York. Also, Mouro I, Colin Y, Cherif-Zahar B, Cartron JP, Le Van Kim C (1993) Molecular genetic basis of the human Rhesus blood group. Nat Genet 5:62–65
13. Carlson EA, Desnick RJ (1979) Mutational mosaicism and genetic counseling in retinoblastoma. Am J Med Genet 4:365–381. Also, Sippel KC, Faioli RE, Smith GD, Schalkoff ME, Sutherland J, Allie BL, Dryja TP (1998) Frequency of somatic and germ line mosaicism in retinoblastoma: implications for genetic counseling. Am J Hum Genet 62:616–619
14. Knudson A (1971) Mutation and cancer: statistical studies of retinoblastoma. Proc Natl Acad Sci 68:820–823
15. McKusick V (1966–1998) Mendelian inheritance in man, 12 editions. Johns Hopkins University Press, Baltimore. It seemed more feasible after 1998 to abandon the costly printed version and switch to a continuously updated on-line version. Thus MIM became OMIM. For each entry disorder there is a date of last revision

Chapter 12
Genetic Studies in Isolated Populations: Victor McKusick's Contributions to Population Genetics

Aravinda Chakravarti

Prologue

Shortly after my arrival at Johns Hopkins in September 2000, as the Director of the McKusick-Nathans Institute of Genetic Medicine, Victor McKusick presented me with a copy of his book *Medical Genetic Studies of the Amish: Selected Papers* [1]. This collection of previously published papers, many authored by McKusick, his collaborators, and his trainees, was assembled with commentary and published by McKusick in 1978. The total body of research represents a broad view of the contributions of the Old Order Amish communities to human medical genetics. McKusick was keenly aware of my prior work in population genetics and complex disease, including among the Old Order Mennonites of Pennsylvania and Ohio, and presented me the book wanting me to be more educated on the genesis and history of these genetic studies. I surmised that he also wanted me to continue these types of studies through the newly founded Institute but now with greater molecular depth in the age of the human genome sequence [2, 3].

McKusick's interest in the Amish arose in 1962 as a "fringe benefit" of university committee work on advising the *Johns Hopkins University Press* [1]. By this time, McKusick had successfully made the transition from cardiologist to geneticist, institutionalized medical genetics as a distinct clinical discipline in the Moore Clinic at Johns Hopkins, established a PhD program in human genetics in the School of Medicine, and started the still-ongoing "Bar Harbor Short Course" [4].

A. Chakravarti, Ph.D. (✉)
Center for Complex Disease Genomics, McKusick-Nathans Institute of Genetic Medicine, Johns Hopkins University School of Medicine, 733 N. Broadway, BRB Suite 579, Baltimore, MD 21205, USA
e-mail: aravinda@jhmi.edu

K.R. Dronamraju and C.A. Francomano (eds.), *Victor McKusick and the History of Medical Genetics*, DOI 10.1007/978-1-4614-1677-7_12,

Importantly, he had gathered around him a remarkably talented group of trainees and junior colleagues and with them created a vibrant and dynamic genetics center in Baltimore, Maryland. Through their cumulative work, McKusick had extensively studied numerous X-linked recessive and autosomal dominant phenotypes but recognized that autosomal recessives were harder to identify and prove in most outbred populations unless they were exceedingly common. Thus, when the *Press* received a manuscript entitled *Amish Society* from John Hostetler [5, 6], McKusick immediately recognized the book's value for his own future genetics work [7]. The Amish, being an essentially closed group with large families who maintained careful genealogical records and were descended from a limited set of founder immigrants, were, McKusick realized, an ideal human community for studying recessive phenotypes. The largest Amish population was resident in Pennsylvania very close to Baltimore and was an attractive feature allowing numerous field visits and direct medical examination for genetic studies.

The argument for identifying recessives in the Amish, particularly "new" recessives, was abundantly clear to McKusick even before he had begun his Amish research. He had, somewhat earlier in 1962, come across a "slick-paper, pharmaceutical company 'throw-away'" magazine article by a Lancaster County physician David Krusen who indicated that achondroplasia was quite frequent among the Amish in Pennsylvania [1, 4]. McKusick immediately realized that since true achondroplasia is a highly penetrant autosomal dominant with reduced fitness, it was unlikely to be common among the Amish unless it were a *new recessive* form. Through discussions with his geneticist colleague Bentley Glass in the Department of Biology, he became aware that the high frequency of a deleterious mutation like recessive dwarfism was likely the result of genetic drift in this isolated community.

This fundamental genetic insight, that isolated inbred communities such as the Amish would reveal numerous new recessive phenotypes that were rare everywhere else but whose mutations had increased in frequency by genetic drift due to the population's unique structure and history (a founder effect), drove McKusick into intensive medical genetic studies among the Amish. The rest is medical genetics history. Nevertheless, one may ask, what has been the legacy of McKusick's insight and his pursuit of this view? In my opinion, there have been three. First, there are the Amish studies themselves, both by him and others outside Johns Hopkins, that have made unique contributions to medical and human genetics [1, 8]. Second, McKusick demonstrated how genetic studies of an exceptional population could enrich and *uniquely* inform human and population genetics [1, 4]. This, of course, is a recurrent and fundamental genetic theme ("Treasure your exceptions" [9]) but applied to populations not individual phenotypes. The value of genetic studies of "isolated" populations is now widely accepted, emulated, and perhaps even oversold for the analysis of complex traits and diseases [10]. Third, the early population studies of the Amish illuminated aspects of human demography and population structure that critically affect how genes can spread through a human population [11]. This last aspect is seldom appreciated, has important implications for future genetics research, and, I predict, will gain in importance.

The Amish Studies

The ostensible reason for McKusick's interest in the Amish was the high frequency of dwarfism reported by David Krusen that he immediately recognized had to be of a new recessive form. His earliest field visits to Amish homes for examining entire families in Lancaster and surrounding counties in Pennsylvania, with David Krusen and John Hostetler, identified many sibships with a previously described rare chondrodystrophy called Ellis-van Creveld syndrome (EvC) that he extensively characterized and described in 1964 [12]. This publication is noteworthy in its breadth for not only investigating the phenotypic variability across 52 affected individuals but also exploring EvC's natural history, genetic segregation as a recessive, inheritance from a single common ancestor Samuel King (Koenig) who immigrated to Pennsylvania in 1744 and estimating the disease incidence (5 per 1,000 live births). The authors were acutely aware of the genetic consequences of two other inferences they made: all Amish cases of EvC were due to a single mutation demonstrating the phenotypic variability that can arise from a single mutation and that the high disease incidence translated into ~13% of the Lancaster County Amish being heterozygotes for the mutation (a founder effect). The *EVC* gene, which when mutated leads to the syndrome, was positionally cloned in 2000 and shown to harbor a splice mutation (IVS13+5G>T) in the Amish [13]. The gene encodes an early developmentally regulated protein that promotes chondrocyte proliferation, hypertrophy, and osteoblast differentiation [14]. Of course, EvC was not the only known Mendelian disorder identified among the Amish; McKusick's studies identified about a dozen other entities [1] whereas subsequent investigators have added more recessive disorders to this list [8].

The more important reason for pursuing genetic studies in an isolate was McKusick's hope of identifying many *new* genetic entities, primarily recessives. Very early on in his quest, in 1965, he hit pay dirt with the identification of a second recessive type of chondrodysplasia (dwarfism) that he termed cartilage hair hypoplasia (CHH), a phenotype that manifested additional immunological defects as well [15]. In yet another comprehensive publication, he summarized the clinical and genetic findings from examination of 77 individuals exploring features of the natural history of CHH, its genetic segregation as a recessive but with incomplete penetrance, inheritance from common ancestors in the 1690s predating their immigration to America, and estimating the disease incidence (1–2 per 1,000 live births). Once again, McKusick was aware of the great phenotypic variability that could result from a single Amish mutation and that the high disease incidence likely occurred through a founder effect through which ~10% of the Amish were heterozygotes for the CHH mutation. The *CHH* gene was identified by positional cloning in 2001 and shown to harbor a common mutation (70A>G) in the RNA component of mitochondrial RNA processing ribonuclease (*RMRP*) gene [16]. Over the next 10 years, McKusick and his trainees, with other colleagues studying Amish populations outside Pennsylvania, identified at least a dozen additional novel recessive entities. Once brought to attention, many were then detected in non-Amish populations.

The Amish studies went beyond the practical utility of identifying new recessive disorders and studying the clinical variability of human genetic disorders. Crucially, and under the urging of both Hostetler and Glass and his colleague Abraham Lilienfeld, McKusick and his trainees delved deep into the genetic structure and demography of these populations [1, 11, 17, 18]. And it was this research that illuminated both the distribution of rare mutations across the geographically distributed Amish populations and defined the conditions under which founder effects became probable. These studies demonstrated that the Amish began with perhaps no more than 200 adult ancestors who migrated to the USA between 1727 and 1770, but they expanded to ~45,000 in the early 1960s [5, 11, 17]; today, they number ~245,000 [8]. This great expansion of population size was accompanied by the westward movement of the primary settlers into Ohio and Indiana and then northward into Ontario, Canada. This was followed by further westward migration into numerous states within the USA and into Mexico. This large 1,000-fold increase in size over ten or more generations from a limited set of founders had two consequences: each new community was serially founded from a preexisting one that further reduced genetic variation, and the Amish practiced consanguinity [19]. These features were significantly evident in studies of family surnames and blood group alleles with each geographic community having its own repertoire of common surnames and alleles [17–19]. These features of population movement and growth gave ample opportunity for random genetic drift and particularly the impressive increases in the allele frequencies of recessive mutations that are rare almost everywhere else. A corollary to this is the rapid disappearance of numerous rare alleles, including deleterious mutations, which all large populations can harbor. Consequently, these Amish studies represented some of the first concrete evolution examples of human in action.

As McKusick noted early in his work, Amish society had several features that were beneficial to human genetics research: (1) they are a very well defined group, (2) it is an essentially closed population, (3) their genealogic records tracing individuals to their immigrant founders are extensive, (4) undernutrition and infectious disease are rare and do not confuse genetic interpretations of disease, (5) their standards of medical care are high and equal across members, (6) socioeconomic uniformity reduced phenotypic variability, (7) family sizes are large (average is seven live births) with low rates or nonexistent illegitimate births, (8) average levels of consanguinity is high, (9) they are agrarian and largely immobile, and (10) they are well informed of the status of their community members [11, 17]. Moreover, the Amish were not "one" population but multiple demes serially derived one from the other so that each deme was useful for the identification of different rare recessive traits that expanded by founder effects. Importantly, the Amish were surprisingly open to genetic investigations that McKusick surmised was due to his (and others') approaching them as physicians [4]. There is no doubt that the Amish have contributed greatly to the study of medical and human genetics through the studies of McKusick. However, many other contemporaries, chiefly Charles Jackson, Herbert Bowman, Richard Juberg, John Opitz, Walter Nance, and Joe Christian, have made significant contributions by studying their own local Amish demes.

In the 1980s, technological advancements made it possible to not only describe and catalog Mendelian entities, as McKusick had earlier begun in *Mendelian*

Inheritance in Man [20, 21], but also to map their locations in the human genome as a prelude to precise gene and mutation identification. The Amish were ideal for mapping recessive disorders since both large pedigrees and genetic homogeneity could be assured, both of which could be limiting in outbred populations. The Amish, through the studies of McKusick and his protégé Clair A. Francomano, contributed to the mapping of the genes for *EVC* [13] and *CHH* [16] and later to that for McKusick-Kauffman syndrome [22]. Two other groups have made extensive use of the Amish for gene mapping and identification: Leslie Biesecker of the National Human Genome Research Institute, National Institutes of Health, Bethesda, MD and Holmes Morton, Clinic for Special Children, Strasburg, PA. The value of the Amish for gene mapping generally, a passion of Victor McKusick's, and one of his major themes for understanding the human genome through its "morbid anatomy," was not lost to McKusick. Indeed, with the help of his junior colleagues at Johns Hopkins, Wilma Bias and Alan F. Scott, McKusick collected ~20 large Amish families each with four grandparents, two parents, and an average of ten offspring, with the expressed purpose of creating a genetic mapping resource similar to the CEPH families (*Centre d'Etude du Polymorphisme* Humain; www.cephb.fr). For detailed studies of the genetic features of the human genome, such as the genome-wide distribution of segregation, recombination, and mutation, these pedigrees remain invaluable today.

Isolated communities like the Amish were of particular value for understanding the phenotypic consequences of variation in any gene because of the limited number of rare alleles from drift. In other words, the existence of many sequence-identical variant heterozygotes and homozygotes meant that phenotypic variation from a single genotype could be better quantified, and genetic modifiers identified with greater precision. Many Amish studies covering blood group markers, chromosomal variants, and common diseases were performed with this objective and are well described in *Medical Genetic Studies of the Amish: Selected Papers* [1]. However, the study of genetic modifiers has lagged behind. Indeed, the Amish represent a unique scenario to investigate whether the incomplete penetrance of the *RMRP* 70A>G CHH mutation is due to a genetic modifier or not; if so, the "rescue" pathway would be important to understand for therapeutic purposes. A particular recent view is that common disorders of complex etiology would be genetically simpler and more tractable in isolated groups like the Amish [10]. Whether this is generally true or not remains to be demonstrated, but its success must depend on the higher frequency of otherwise rare recessive variants.

Studies of Other Isolated Populations

Medical genetic studies of the Amish were a forerunner to numerous investigations of other "isolated" populations and their particular suite of otherwise rare recessive disorders. Indeed, the definition of isolated is considerably variable and inconsistent and runs the gamut from studies of small, undifferentiated communities (Arab villages within Israel, e.g., Ein Mahel), to highly specific small (Old Order Mennonites,

Hutterites) or large (Ashkenazim) religious communities, to specific geographic regions with restricted ancestry (Saguenay–Lac-Saint-Jean, Quebec, Canada), and even to entire nations (Finland). There is no doubt that hundreds of communities across the world can fit within this broad definition. The common feature to the claim of genetic isolation is, analogous to the Amish, descent from a limited set of founders, rapid population growth, moderate to high levels of consanguinity, and large family sizes. Together, these features increase the chance of genetic drift and founder effects from rare alleles, but generally, many of these other populations do not have the other crucial advantages of the Amish, such as low rates of mobility, immigration, illegitimate births, and a uniform lifestyle and environment across all members. I will discuss two specific examples to show how additional studies among the Amish and Mennonites have proceeded and the more contemporary and extensive genetic studies in Finland.

The recognition of new clinical entities and gene mapping in the Amish continued throughout the 1980s and 1990s even when McKusick was not himself the protagonist. However, without his direct involvement, genetic studies of the Amish became less intense in Pennsylvania for a while. During this time, my own involvement with these communities began with genetic studies of Hirschsprung disease. By 1988, my laboratory was convinced that we could identify genes for the multifactorial birth defect Hirschsprung aganglionosis, despite its complex inheritance, since both dominant and recessive major genes could be postulated based on segregation [23, 24]. Therefore, we were on the hunt for large pedigrees suitable for genetic mapping. Based on discussions with William Sieber, a senior surgeon colleague at the University of Pittsburgh, Pennsylvania, where I was a junior faculty member, who had repaired a number of Mennonite children with Hirschsprung disease at The Children's Hospital, Pittsburgh, and one of his colleagues at the Hershey Medical Center who had a number of such children under his care [25], I became convinced that these Mennonite cases likely segregated a major recessive mutation.

Our preliminary enquiries pointed to the existence of a common mutation for Hirschsprung disease among the Old Order Mennonites of Lancaster and Berks counties in Pennsylvania. Consequently, with the use of a "circle letter," we organized a 1990 family reunion in Kutztown, Pennsylvania, to obtain family histories and examine over 200 Mennonites and a few Amish families. I had the collaboration of two medical genetics colleagues, namely, Kenneth Garver of the University of Pittsburgh and Roger Ladda of the Hershey Medical Center. Holmes Morton also appeared at this reunion since he was studying maple syrup urine disease (MSUD), in high frequency among the Old Order Mennonites, in some of the same families. We examined families that had at least one child with Hirschsprung disease; it appeared that many of these sibships had segregation of sensorineural deafness, hypopigmentation, and heterochromia irides; one rare Amish sibship had both Hirschsprung disease and cartilage hair hypoplasia (CHH). Our subsequent studies demonstrated a remarkably extensive kindred which demonstrated a high frequency of Hirschsprung disease that could be traced back to a few common immigrant ancestors ~10 generations ago, non-Mendelian inheritance, sex difference in expression but largely the effects of at least one major gene that was a susceptibility factor for aganglionosis, deafness and pigmentary anomalies [26].

Our research on the Mennonite Hirschsprung disease kindred led us through all of the lessons that McKusick himself had learnt and pointed out to others [11]. We discovered that the disorder segregated not so much within a kindred as through the entire population, the presumptive mutation(s) was common not only in the Pennsylvania Mennonites but in other Old Order communities in Ohio and Ontario and that a founder effect was most probably the reason for the high frequency of the disorder. However, atypical of other recessive traits, the disorder had a segregation ratio of only 8% and showed a marked sex difference and a curious pattern of pigmentary anomalies and deafness both as an isolated entity and in combination with aganglionosis. Nevertheless, the recognition of common ancestry and the likelihood of a founder effect made genetic mapping by recognizing genomic segments inherited in common between affected individuals ("identity-by-descent" mapping) attractive even if the trait did not Mendelize. We were successful in identifying a primary locus for Hirschsprung disease on human chromosome 13 and a putative modifier on human chromosome 21 by 1994 [26]. Shortly thereafter, we were also successful in positionally identifying the endothelin receptor type B (*EDNRB*) gene as the major susceptibility gene which explained most features of the disorder including the apparent pleiotropy as due to tissue specific differences in the mutation's effect in homozygotes versus heterozygotes [27].

Much of this work relied on the efforts of one remarkable graduate student, Erik Puffenberger. Erik, who is from Pennsylvania, of Mennonite ancestry, and with a deep abiding interest in the "Plain" peoples, genealogies, and human genetics, went to work as a research technician at Johns Hopkins after completing his undergraduate degree from Swarthmore College in the outskirts of Philadelphia. He also became McKusick's gofer and thereby came into intimate contact with McKusick, *Mendelian Inheritance in Man* and gene mapping. Desirous to complete a doctoral degree in gene mapping of the Amish and Mennonite disorders, he was recommended to me by my collaborators in Hopkins (I was at the University of Pittsburgh then). This was a match made in heaven! There is no doubt that Erik was the driving force of gene mapping in the Old Order Mennonites for Hirschsprung disease in my laboratory since he nearly single-handedly completed the relevant genealogy, identified and sampled additional sibships, and performed the genome screens that led to mapping and then gene identification. Much of the impetus and success for further gene mapping, gene identification, and population genetics of founder effect mutations in the Amish and Mennonites since then have been spearheaded by Erik but in a different capacity.

A parallel development has assured that these types of genetic studies will have a home, at least for Amish and Mennonite studies. In 1989, Holmes Morton, a pediatrician, and his wife, Caroline Morton, established a small nonprofit clinic called the Clinic for Special Children (www.clinicforspecialchildren.org/) in Strasburg, Pennsylvania, and dedicated to the care and treatment of Amish and Mennonite children with genetic disorders. Importantly, the clinic represents one of the best examples of community genetics where both clinical care and research are interwoven. After completing his doctoral degree with me, Erik Puffenberger returned to Lancaster County and to the Clinic for Special Children where he became laboratory director. Here, Erik has used remarkably modest resources to be immensely successful in using identity-by-descent mapping to map and identify the genes and mutations for more than 25 disorders [8]. Moreover, as required for their clinical management of their patients, Erik and his colleagues have made available genetic testing for >37 Amish and >29 Mennonite disorders common in Pennsylvania, Ohio, and Maryland. Today, when DNA sequencing is both accessible and affordable, one can envision that all Amish and Mennonite genetic disorders can be identified even without mapping and I expect that the Clinic will play a leading role. Interestingly, an analogous clinic, the *Das Deutsch Center for Special Needs Children*, has been created for the service of the Ohio Amish and Mennonite communities. These two clinics fulfill many aspects of McKusick's vision that basic investigations in genetics will ultimately benefit patients who contribute and collaborate on these efforts.

While McKusick was engaging in his Amish studies, there was a similar movement underway in Finland. In the 1950s, physicians at the Children's Hospital, University of Helsinki, noted the high frequency of a form of fatal congenital nephrosis; it was also familial. Although a nationwide study was begun to identify the environmental factor presumed to be involved, further studies definitively established it as an autosomal recessive disorder [28]. Over the next decade, numerous Finnish investigators defined many additional clinical traits that were rare elsewhere but exceedingly common in Finland. The vast majority of these were also familial and autosomal recessive. They comprise what has been termed the Finnish Disease Heritage (FDH), a collection of rare diseases that are overrepresented in Finland, a feature long recognized by Finns. The explicit recognition of the FDH was by Perheentupa in 1972 [29], but its genetic causes, arising from national and regional isolation, were first explicitly described by Norio in 1981 [30]. Here again, regional isolation followed by rapid size expansion with the resultant consanguinity and large family size favored genetic drift of rare mutant alleles with an unusual distribution of autosomal recessives. Some autosomal recessive disorders were absent while others were overrepresented, this overrepresentation varying by region within Finland. Geneticists and physicians within Finland have been remarkably adept and successful in identifying the gene and mutations and studying the population distribution of more than 20 diseases as a part of the FDH [30, 31]. I have no doubt that many additional FDH diseases will be identified and solved in the future.

The Amish-Mennonite and Finnish studies are the most systematic genetic investigations of genetic diseases in isolated communities, so it is instructive to compare and contrast them. Both of these populations represent historical cases where genetic isolation, geography in Finland and religion in the Amish-Mennonite, of a small group of founders enhanced the genetic drift of rare mutations and led to remarkable founder effects. Consanguinity is not necessary in these circumstances since homozygotes segregate through both parents who are ancestrally related thorough descent from multiple common ancestors in the near past (10–20 generations ago).

However, consanguinity is an almost constant feature of small populations and will only enhance the appearance of recessives. Interestingly, McKusick and Perheentupa, both of whom were physicians, came to similar conclusions but from very different perspectives. Perheentupa [29] noticed the burden of *unique Finnish diseases* and sought a common explanation while McKusick [17] recognized the *uniqueness of Amish society* for genetic studies and identified new genetic diseases as a consequence. In retrospect, it appears that McKusick perspective was more genetically motivated and general.

Epilogue

The classical Amish studies, and no doubt the Finnish studies of recent vintage, have made a lasting impression on human genetics. First, the importance of founder effects in the identification of recessive diseases is widely recognized. Indeed, it is likely that to date, when mapping has been a prerequisite to gene identification, we have largely uncovered those recessive disorders whose alleles have been buoyed up by genetic drift in some population or the other or consanguinity has allowed for homozygosis of a rare mutation. Second, human geneticists routinely search for isolated communities where the chance of detecting recessives is enhanced. There are hundreds, if not thousands, of such communities around the world who have been isolated by geography, culture, or religion, and many such communities are as large as the Amish demes. Third, now that identity-by-descent can be effectively distinguished by a high marker density or sequencing the human genome [2, 3], even limited isolation and smaller degrees of founder effects can be recognized. Consequently, as the more recent Amish-Mennonite studies have shown [8], the impact of these populations on human genetics is likely to increase further.

Much of the focus on isolated populations has been from the disease perspective. However, it has become amply clear that in order to understand the spread of disease mutations by drift, we require population genetics studies to be intricately entwined with medical genetic studies [11, 12, 17, 18, 26, 27, 30]. Indeed, contemporary identity-by-descent mapping requires population genetic models [27]. Consequently, it is not hard to predict that despite their rarity, isolated populations will play a significant role in population genetics distinct from human medical genetics.

A lasting question is whether "isolated populations" are rare exceptions [9] or more central to human genetics. Genome-wide data from multiple contemporary human populations have clarified that all extant humans are descendants of relatively small populations, inside or "out-of-Africa," with effective sizes smaller than 10,000 [32]. Even large subcontinental populations, such as that of India, are not homogeneous but composed of innumerable identifiable groups each of which have much smaller effective sizes (1,000 or smaller) despite their current numerical strength [33]. Thus, the contribution of genetic drift and founder effects to human evolution generally, and all human diseases specifically, may be far greater than we currently envision particularly with recent evidence suggesting multiple serial founders for the human population [34]. With the sequencing of human genomes and the identification of "rare" genetic variants within genes [35], it is likely that most functional genetic variants are geographically localized and have been subject to founder effects. If so, the Amish studies, where mutations are generally localized to demes, may be a "model" for human trait and disease evolution and go far beyond what McKusick had envisioned.

References and Notes

1. McKusick VA (1978) Medical genetic studies of the Amish. Johns Hopkins University Press, Baltimore
2. International Human Genome Sequencing Consortium (2001) Initial sequencing and analysis of the human genome. Nature 409:860–921
3. Venter JC, Adams MD, Myers EW, Li PW, Mural RJ et al (2001) The sequence of the human genome. Science 291:1304–1351
4. McKusick VA (2006) A 60-year tale of spots, maps and genes. Annu Rev Genomics Hum Genet 7:1–27
5. Hostetler JA (1963) Amish society. Johns Hopkins University Press, Baltimore
6. The *Johns Hopkins University Press* has published numerous books on Amish and Hutterite culture and society owing to the huge success of Hostetler's *Amish Society*, first published in 1963 with a fourth edition in 1993, which became one of its best selling volumes
7. John Andrew Hostetler (1918–2001), born into an Old Order Amish family and who later joined the Mennonite church, was an educator and leading sociologist scholar of Amish and Hutterite society with whom McKusick had a lifelong collaboration and friendship (http://en.wikipedia.org/wiki/John_A._Hostetler). Hostetler introduced McKusick to the Amish, taught him much about the Amish way and view of life, and made the medical genetic studies possible
8. Strauss KA, Puffenberger EG (2009) Genetics, medicine and the plain people. Annu Rev Genomics Hum Genet 10:513–536
9. "If I may throw out a word of counsel to beginners, it is: Treasure your exceptions! When there are none, the work gets so dull that no one cares to carry it further. Keep them always uncovered and in sight. Exceptions are like the rough brickwork of a growing building which tells that there is more to come and shows where the next construction is to be." William Bateson (1908) The method and scope of genetics. Cambridge University Press, Cambridge
10. Heutink P, Oostra BA (2002) Gene finding in genetically isolated populations. Hum Mol Genet 11:2507–2515
11. McKusick VA, Hostetler JA, Egeland JA, Eldridge R (1964) The distribution of certain genes in the Old Order Amish. Cold Spring Harb Symp Quant Biol 29:99–114
12. McKusick VA, Egeland JA, Eldridge R, Krusen DE (1964) Dwarfism in the Amish. I. Ellis-van Creveld syndrome. Bull Johns Hopkins Hosp 115:306–336
13. Ruiz-Perez VL, Ide SE, Strom TM, Lorenz B, Wilson D et al (2000) Mutations in a new gene in Ellis-van Creveld syndrome and Weyers acrodental dysostosis. Nat Genet 24:283–286
14. Pacheco M, Valencia M, Caparrós-Martín JA, Mulero F, Goodship JA, Ruiz-Perez VL (2012) Evc works in chondrocytes and osteoblasts to regulate multiple aspects of growth plate development in the appendicular skeleton and cranial base. Bone 50:28–41
15. McKusick VA, Eldridge R, Hostetler JA, Ruangwit U, Egeland JA (1965) Dwarfism in the Amish. II. Cartilage-hair hypolasia. Bull Johns Hopkins Hosp 116:285–326
16. Ridanpaa M, van Eenennaam H, Pelin K, Chadwick R, Johnson C et al (2001) Mutations in the RNA component of RNase MRP cause a pleiotropic human disease, cartilage-hair hypoplasia. Cell 104:195–203

17. McKusick VA, Hostetler JA, Egeland JA (1964) Genetic studies of the Amish. Background and potentialities. Bull Johns Hopkins Hosp 115:203–222
18. Cross HE, McKusick VA (1970) Amish demography. Social Biol 17:83–101
19. Jackson CE, Symon WE, Pruden EL, Kaehr IM, Mann JD (1968) Consanguinity and blood group distribution in an Amish isolate. Am J Hum Genet 20:522–527
20. McKusick VA (1964) On the X chromosome of man. AIBS, Washington, DC; McKusick VA (1964) Mendelian inheritance in man: catalogs of autosomal dominant, autosomal recessive and X-linked phenotypes, 1st edn. Johns Hopkins University Press, Baltimore, 525 pp
21. OMIM is the internet (online) version of Mendelian Inheritance in Man which was made available starting in 1987 from the Welch Medical Library at Johns Hopkins and from December 1995 from the National Center for Biotechnology Information (NCBI) of the National Library of Medicine (www.ncbi.nlm.nih.gov/omim). It is, once again, offered through Johns Hopkins at www.omim.org
22. Stone DL, Agarwala R, Schäffer AA, Weber JL, Vaske D et al (1998) Genetic and physical mapping of the McKusick-Kaufman syndrome. Hum Mol Genet 7:475–481
23. Badner JA, Chakravarti A (1990) Waardenburg syndrome and Hirschsprung disease: evidence for pleiotropic effects of a single dominant gene. Am J Med Genet 35:100–104
24. Badner JA, Sieber W, Garver KL, Chakravarti A (1990) A genetic study of Hirschsprung disease. Am J Hum Genet 46:568–580
25. Cohen IT, Gadd MA (1982) Hirschsprung's disease in a kindred: a possible clue to the genetics of the disease. J Pediatr Surg 17:632–634
26. Puffenberger EG, Kauffman ER, Bolk S, Matise TC, Washington SS et al (1994) Identity-by-descent and association mapping of a recessive gene for Hirschsprung disease on human chromosome 13q22. Hum Mol Genet 3:1217–1225
27. Puffenberger EG, Hosoda K, Washington SS, Nakao K, deWit D, Yanagisawa M, Chakravarti A (1994) A missense mutation of the endothelin-B receptor gene in multigenic Hirschsprung's disease. Cell 79:1257–1266
28. Norio R (1966) Heredity in the congenital nephritic syndrome: a genetic study of 57 Finnish families with a review of reported cases. Ann Paediatr Fenn 12(Suppl 27):1–94
29. Perheentupa J (1972) Suomalainen tautiperintö (symposium on inherited disease in Finland). Duodecim 88:1–166
30. Norio R (2003) Finnish disease heritage. I: Characteristics, causes, background. Hum Genet 112:441–456
31. Norio R (1981) Diseases of Finland and Scandinavia. In: Rothschild H (ed) Biocultural aspects of disease. Academic, New York, pp 359–415
32. Chakravarti A (2009) Being human: kinship: race relations. Nature 457:380–381
33. Reich D, Thangaraj K, Patterson N, Price AL, Singh L (2009) Reconstructing Indian population history. Nature 461:489–494
34. DeGiorgio M, Jakobsson M, Rosenberg NA (2009) Explaining worldwide patterns of human genetic variation using a coalescent-based serial founder model of migration outward from Africa. Proc Natl Acad Sci USA 106:16057–16062
35. The 1000 Genomes Project Consortium (2010) A map of human genome variation from population-scale sequencing. Nature 467:1061–1073

Chapter 13
Victor A. McKusick and Medical Genetics Among the Amish

Clair A. Francomano

In 1962, two events led Dr. Victor McKusick to embark on what became a lifelong connection to the Amish people. His work among the Amish sparked a whole new branch of genetics, namely, that of the genetics of inbred populations, and led to the recognition and characterization of dozens of new Mendelian disorders, blood groups, and chromosomal variants. He and his colleagues initiated studies of blood pressure, diabetes, and cervical carcinoma, demonstrating the potential of the Amish communities for analysis of common disorders. As time went on, other investigators came to recognize that the very characteristics of Amish society that made it an ideal laboratory for the study of Mendelian disorders also lent themselves to the dissection of quantitative and common traits such as hypertension and hyperlipidemia. Along the way, McKusick's interactions with the Amish people gave them additional insight into the risks of recessive disorders and led to some changes in the social mores of the society.

As McKusick himself tells the story in the Preface to his 1978 book Medical Genetic Studies of the Amish [11], his interest in the Amish community was first piqued in 1962 by an article about Dr. David Krusen, a country doctor with a large rural practice among the Amish at the time. The article appeared in a glossy pharmaceutical publication. In it, Dr. Krusen reported that he was seeing many patients with achondroplasia, the most common form of dwarfism, among his Amish patients. McKusick recognized that the population structure of the Amish was more compatible with a higher frequency of recessive disorders, rather than the autosomal dominant achondroplasia. He suspected that the dwarfism Dr. Krusen was seeing was in fact a different kind of dwarfism similar to achondroplasia but clinically and genetically distinct from it.

C.A. Francomano, M.D. (✉)
Harvey Institute for Human Genetics, Greater Baltimore Medical Center,
6701 North Charles Street, Baltimore, MD 21204, USA
e-mail: cfrancomano@gbmc.org

K.R. Dronamraju and C.A. Francomano (eds.), *Victor McKusick and the History of Medical Genetics*, DOI 10.1007/978-1-4614-1677-7_13,

The second event that occurred in 1962 was that the Johns Hopkins University Press received a manuscript from John Hostetler entitled Amish Society [47]. McKusick was on the faculty committee advising the Press at the time. He was impressed by the many features of the Amish society which made it ideal for the study of genetic disorders, particularly autosomal recessive disorders. At that time in his genetics career, he had already embarked on extensive familial studies of a number of autosomal dominant and X-linked traits; the ready availability of a large number of recessive conditions in Lancaster County, Pennsylvania, a community geographically convenient to his hometown of Baltimore, was the spark for immediate action. He embarked upon a collaboration with John Hostetler that endured for many years, and his initial studies in the community included travels with Dr. Krusen to see the families under his care.

Thus began more than four decades of field trips to Amish communities all over the United States. Focusing initially on studies of dwarfism, McKusick, his trainees, and collaborators found two recessive forms of short stature to be relatively common among the Amish. The first, a disorder that had been described previously but was exceedingly rare, was Ellis-van Creveld syndrome, or six-fingered dwarfism [66, 67]. The second, which McKusick named cartilage-hair hypoplasia [1], later came to be known as the McKusick type of metaphyseal chondrodysplasia. This disorder was later found in high frequency among the Finnish population as well and represented the first of many disorders recognized first among the Amish but later found among other populations with similar population structure.

In addition to the early studies of dwarfism among the Amish, the early trips to Lancaster County also led to extensive investigation of blood groups [2], as well as the HLA system and the study of immunogenetics, under the auspices of Dr. Wilma Bias [3, 31, 32, 49, 50, 70, 71]. Other studies systematically approached by the Hopkins genetics teams included mental retardation [4], Down syndrome [5], neurologic disorders, and deafness [6]. McKusick and his colleagues reported on extensive families with previously recognized disorders including ataxia-telangiectasia [7], pyruvate kinase deficiency [8], epidermolysis bullosa, symphalangism, and many others [34, 37, 53, 57, 63]. Studies among the Amish in Holmes County, Ohio, led to the recognition of new neurologic disorders: Troyer and Mast syndromes [9, 10]

In his 1978 book, Medical Genetic Studies of the Amish [11], McKusick cites the work that was being done by others among the Amish communities across the United States, either simultaneous to his initial work or in some cases preceding his studies. These investigators included Charles E. Jackson, Herbert S. bowman, Richard C. Juberg, Wlater E. Nance, Joe C. Christian, and John M. Opitz. McKusick also credits Dr. D Bernard Amos of Duke University for initiating the immunogenetic studies, in part in collaboration with his group at Hopkins. Manuscripts from each of these investigators and many others are included in the 1978 compendium. Medical Genetic Studies in the Amish [11] includes reports describing 16 newly recognized Mendelian disorders as well as 18 previously recognized conditions among the Amish communities in Pennsylvania, Ohio, and Indiana. Reports describing the early immunogenetic studies, investigation of chromosomal variations and aberrations, and analysis of common disorders such as blood pressure, diabetes, and cervical cancer are also included.

In his seminal paper on the population structure of the Amish and the potential for genetic studies among them, McKusick and colleagues [68] describe the societal and demographic features of Amish society that render the community particularly useful for genetic studies. These included:

1. The Old Order Amish represent a well-defined population, of which the membership is carefully documented and unequivocal.
2. The Amish population is closed. Although some young people may choose to leave the community, it is extremely rare for anyone to join the group. Marriage outside the faith is forbidden.
3. The origins of the population are Western European and well known.
4. Extensive genealogical records exist. In 1964, there were at least 60 publications of Amish family records, including the Fisher genealogy, which listed the children and their dates of birth (and death), of more than 3,600 families in Lancaster County at the time. These genealogical records eventually came to be computerized in work done by Dr. Leslie Biesecker and his colleagues at the National Institutes of Health in the 1990s, resulting in the Amish Genealogy Database [59].
5. There is a high standard of living. McKusick intuited that the role played by genetic factors in causing morbidity and mortality would not be masked by nongenetic factors such as malnutrition and infection.
6. There is a relatively high standard of medical care, leading to likely correct identification of specific diagnoses and ascertainment of cases in the early days of disease identification and description.
7. There is a high interest in the state of health and of illness among members of the community. The Budget, a community newspaper, reports illnesses among the people and causes of death for members of the population.
8. There is a high degree of consanguinity as a result of the relatively few founder families who established each of the Amish communities across the United States. This observation was later expanded upon by the work of Dr. Muin Khoury, who elaborated on the epidemiology of genetic disease among the Amish.
9. Rates of nonpaternity were felt to be quite low as a result of the strict regulations of the Amish church.
10. The Amish are clannish and well informed about illness in close relatives. Sara Fisher, an Amish schoolteacher who served as a liaison with McKusick for many years, used to joke that although the Amish did not have telephones, they had an even more effective means of communication: "tell-a-woman."
11. Socioeconomic and occupational circumstances are uniform, further reducing the environmental contribution to variability in the population.
12. There is relative immobility among the Amish because of their use of the horse-and-buggy transportation and commitment to a life of farming.
13. Amish couples tend to have large families with sibships averaging seven to nine in number. Not only do the Amish avoid birth control, they consider children born with congenital malformations to be "special gifts" and feel privileged to

have the opportunity to care for these children with exceptional needs. In light of the geographic immobility of the population, this results in large numbers of relatives and individuals with disorders of interest, readily available for study within a relatively restricted geographic locale.

14. Individuals with congenital malformations and intellectual handicaps are usually not institutionalized but kept at home.
15. There are a number of separate Amish isolates, each with distinct founding populations, making comparisons between the groups not only feasible but relatively easy.

These characteristics of isolated and inbred populations remain as relevant to genetic studies today as they were nearly 50 years ago.

McKusick cited several potential obstacles to utilizing the full potential of the Amish for genetic studies, including the injunction of the Amish faith to "remain separate from the world," ethnic solidarity, and a strong antiscientific attitude, including a reluctance to allow autopsy on deceased family members.

McKusick and his colleagues relied on close collaborations with members of the Amish community to facilitate his entrée into Amish homes. Sara Fisher, an Amish school teacher, was the liaison for researchers from Johns Hopkins for many years, beginning in the mid-1960s and extending well into the 1990s when her health began to fail. Sara was a highly intelligent and intuitive woman with whom I had the pleasure to work for many years, first as a trainee and later as a faculty member in the Department of Medicine. She was proud of her work with Dr. McKusick and would always take us to the Amish homes in Lancaster County by different routes, so we would remain dependent on her guidance in those pre-GPS days. She called this "job insurance." Her middle initial was E., which she joked was "short for eat," because she did love to eat the delicious Amish cuisine to be found in the many restaurants around Lancaster County. In fact, the naming custom among the Amish mandates that a child's middle initial be the initial of the mother's maiden name; Sara's mother had been an Esh before she married. McKusick often brought visiting dignitaries to Lancaster for a trip to see the beautiful Amish countryside, meet Sara Fisher and see her one-room schoolhouse, and learn about the unique way of life among the Amish people. These visits were invariably punctuated by meals at Sara's favorite restaurant, the Good 'N Plenty, which provided family style meals of enormous proportions.

McKusick and colleagues, in their "Background and Potentialities" paper [68], noted three types of genetically significant information obtainable from studying a population such as the Amish:

1. "New" recessively inherited disorders can be discovered. McKusick viewed this endeavor as a means of understanding the normal genetic constitution of humans. He posited that "catalogs of simply inherited genetic traits are like photographic negatives from which a positive picture of the normal genetic constitution of man can be constructed."
2. Evidence supporting the recessive inheritance of extremely rare phenotypes may be found.

3. By studying many cases of a rare recessive disorder in an endogamous isolate population, a much fuller picture of the characteristics and variability within a disorder may be put forward.
4. By tracing rare recessive diseases back to a single common ancestral couple, it is theoretically possible to estimate the fraction of the genetic load in specific ancestors due to specific recessive genes. McKusick et al. also noted that the inheritance of specific chromosomal anomalies is of interest in inbred populations, particularly bearing in mind that there may be alleles which in the homozygous state predispose to nondisjunction; they hypothesize that the existence of such genes in man may result in an increased frequency of chromosomal nondisjunction among families of inbred populations. They also note that by studying the frequencies of blood groups and other marker traits, investigators may ascertain the degree of distinctness of sub-isolates and the operation of genetic drift.

Amish Demographics

Cross and McKusick [12], and later McKusick [64], reported on the demography of the Amish population in Holmes County, Ohio. They note several interesting features of the population. A comparison of age pyramids of Amish and their rural neighbors supports the perceived social and cultural isolation of the Amish community. They found that Amish women in this population had a median number of 6.66 live births, leading to a potential doubling of the population every 23 years. Moreover, for reasons that Cross and McKusick were unable to ascertain, they found the male to female sex ratio of Amish over the age of 55 to be 1.216, much higher than that seen in the general United States population. The finding that men tend to outlive their wives in this population was replicated in the Hutterites and in a later study of Amish in Indiana by Juberg [101]. Eaton and Mayer [100] raised the possibility that multiparity might contribute adversely to longevity among women in these populations, and Juberg's data from 1966 seem to support this hypothesis, as the oldest women had the lowest mean number of offspring in that study. In the 1980's, Khoury et al. [54–57] reported on further studies of inbreeding and pre-reproductive mortality among the Old Order Amish.

Cross and McKusick's data [12] also suggested genetic contribution to twinning in the Holmes County Amish community, as 9.9% of the families with twin births accounted for 17.5% of twin births.

Early on, McKusick recognized that the distinct distribution of names in the various Amish communities (called demes) across the United States reflected the unique founder couples for each community and hence their novel genetic constitution [65, 69]. Subsequent studies reported by Strauss and Puffenberger [13] demonstrate that mitochondrial disease haplotypes, representing matrilineal inheritance, demonstrate similar drift to that of the surnames, which are paternally inherited, over the previous 12–14 generations.

Dwarfism in the Amish

Two papers, one by McKusick et al. in 1964 [66] and a second by Murdoch and Walker in 1969 [102], described the Ellis-van Creveld syndrome in the Lancaster County, PA, Amish population. McKusick updated those reports in a commentary in his 1978 compendium, Medical Genetic Studies of the Amish [11]. By that time, a total of 40 affected sibships and 82 affected individuals had been documented. All 80 parents traced their ancestry to Samuel King and his wife. The phenotype of dwarfism in persons affected by the Ellis-van Creveld syndrome (EvC) includes mesomelic and acromelic shortening of the limbs (affecting the mid-to-distal region of the limbs), postaxial polydactyly, dysplastic finger- and toenails, natal teeth, tethered frenulum of the upper lip, and congenital heart disease, including atrial septal defect in approximately 50% of affected persons. In his 1978 commentary [11], McKusick notes that the differential diagnosis of EvC must include Jeune asphyxiating thoracic dystrophy, which also has polydactyly as a feature in some affected persons, but is more likely to have polydactyly affecting the feet, and does not manifest nail dysplasia.

It was to be 36 years from the initial report of EvC among the Amish until the gene causing the condition was found [62, 105] and another 10 years before its pathogenesis began to be understood. Two genes, arranged in a head-to-head configuration, were found located on chromosome 4p16 and named EVC1 and EVC2 [106]. However, mutations in these two genes account for only about 2/3 of cases [109], suggesting that at least a third locus remains to be found. The Evc protein has been localized to the base of the primary cilium of chondrocytes, and defective Ihh signaling was observed in proliferating chondrocytes of Evc-null mice [107]. Currently, Ellis-van Creveld syndrome is considered to be a ciliopathy [14, 45].

The syndrome that later came to be known as the McKusick type of metaphyseal chondrodysplasia was reported by McKusick et al. as Cartilage-Hair Hypoplasia in 1965 [1, 67]. The early paper reported short-limbed dwarfism (accounting for the superficial resemblance to achondroplasia), normal head circumference (as opposed to the macrocephaly observed in achondroplasia), and short, loose-jointed hands and feet with short finger- and toenails. Excessive length of the fibula relative to the tibia was noted. Other manifestations included sparse, fine hair that breaks easily; in the extreme, affected persons may be nearly bald. Relatively light color of the hair is an inconsistent feature. Gastrointestinal manifestations observed among some affected persons included Hirschsprung's disease. Affected persons seemed to be unusually susceptible to viral infections and malignancy, particularly lymphoma. Indeed a subsequent report documented chronic neutropenia and abnormal cellular immunity among affected persons (Lux et al. 1970). Multiple studies in the ensuing 40 years have expanded and clarified the nature of the immune defect in this population of patients [78, 92].

Nevanlinna [103] reported on CHH in the Finnish population and reviewed 30 cases in 1977 [104]. McKusick's commentary in 1978 [11] (p. 279) notes the parallels between Finnish population and structure and posits "independent but identical mutations that acquired a relatively high gene frequency in each population because of similar founder-drift factors." This remarkably prescient statement was born out by later studies identifying mutations in the RMRP gene, located on chromosome 9p13-12, as the cause of CHH [91, 108], with the major mutation causing CHH in the Amish being the same as that seen in most Finnish cases [15].

Other Mapping Studies

In the 1980s and 1990s, mapping studies with microsatellite markers enabled gene identification of multiple phenotypes in the Amish and related populations including the Mennonites [43]. Genetic conditions elucidated by microsatellite mapping included metabolic disorders such as propionic acidemia and intrahepatic cholestasis, neuromuscular diseases including recessive limb-girdle muscular dystrophy and Troyer syndrome, congenital malformation syndromes including congenital microcephaly and McKusick-Kaufman syndrome, as well as multigenic Hirschsprung's disease.

The Clinic for Special Children was founded by Dr. Holmes Morton and his wife Caroline in 1989. Located in Lancaster County, the clinic is devoted to providing affordable, accessible care to nearby families, as well as to the investigation of genetic disorders in the community. Driven by the clinical needs of his patients, Morton and his colleagues pioneered methods for diagnosis and early management of glutaric aciduria type 1 [16, 84] and maple syrup urine disease [17, 89, 90].

The advent of large-scale SNP genotyping has enabled Morton and his colleagues to locate and find more than 75 different mutations among the Amish and Mennonite people. As Strauss and Puffenberger relate in their 2009 review [13], they have used candidate gene analysis, candidate gene localization, and finally disease gene mapping, by comparing homozygous blocks among multiple affected persons, to find causative genes. As of 2009, over 40 different genetic mapping studies had been initiated; 14 had yielded the pathologic allele and 10 others were mapped to a chromosomal region, but without definitive identification of the causative gene. A complete list of 37 Amish and 29 Mennonite disorders understood at the molecular level as of 2009 is included in Strauss and Puffenberger's review [13].

Another investigator who followed in McKusick's footsteps among the Amish of Lancaster County was Alan Schuldiner, a genetic epidemiologist whose efforts were focused on the genetics of complex traits. He confirmed McKusick's impression that the societal features which enabled the study of Mendelian traits would also facilitate the study of traits such as coronary artery disease [81, 82], arterial stiffness [18], lipid profiles [19], hypertension [20], and many others [21–28, 83, 85].

These prolific results, and those of many others who have followed in his footsteps [26–30, 33, 35, 36, 38, 41, 42, 44, 46, 48, 51, 52, 58, 61, 62, 72–77, 79, 80, 86–88, 93–99] among isolated populations too numerous to mention throughout the world, amply justify McKusick's early and prescient enthusiasm for the study of genetic disorders in the Amish. Through the identification of multiple disease genes and risk alleles, these studies have indeed brought to fruition his prediction that "catalogs of simply inherited genetic traits are like photographic negatives from which a positive picture of the normal genetic constitution of man can be constructed."

References

1. McKusick VA, Eldridge R, Hostetler JA, Ruangwit U, Egeland JA (1965) Dwarfism in the Amish. II. Cartilage-hair hypoplasia. Bull Johns Hopkins Hosp 116:285–326
2. Bias WB, Light-Orr JK, Krevans JR et al (1969) The Stoltzfus blood group, a new polymorphism in man. Am J Hum Genet 21:552–558
3. Hsu SH, Pollard MK, Chan MM, Bias WB (1981) Genetic heterogeneity of DR4 in the old order Amish and two new HLA-D specificities. Hum Immunol 2:165–172
4. Gale AN, Lacassie Y, Rogers JG, Levin LS, McKusick VA (1977) Two "new" autosomal recessive mental retardation syndromes observed among the Amish. Birth Defects Orig Artic Ser 13:127–138
5. Juberg RC, Davis LM (1970) Etiology of nondisjunction: lack of evidence for genetic control. Cytogenetics 9:284–293
6. Eldridge R, Berlin CI, Money JW, McKusick VA (1968) Cochlear deafness, myopia, and intellectual impairment in an Amish family. Arch Otolaryngol 88:49–54
7. McKusick VA, Cross HE (1966) Ataxia-telangiectasia and Swiss-type agammaglobulinemia. Two genetic disorders of the immune mechanism in related Amish sibships. JAMA 195:739–745
8. Bowman HS, McKusick VA, Dronamraju KR (1965) Pyruvate kinase deficient hemolytic anemia in an Amish isolate. Am J Hum Genet 17:1–8
9. Cross HE, McKusick VA (1967a) The Troyer syndrome. A recessive form of spastic paraplegia with distal muscle wasting. Arch Neurol 16:473–485
10. Cross HE, McKusick VA (1967b) The Mast syndrome. A recessively inherited form of presenile dementia with motor disturbances. Arch Neurol 16:1–13
11. McKusick VA (1978) Medical genetic studies of the Amish: selected papers. Johns Hopkins University Press, Baltimore
12. Cross HE, McKusick VA (1970) Amish demography. Soc Biol 17:83–101
13. Strauss KA, Puffenberger EG (2009) Genetics, medicine, and the plain people. Annu Rev Genomics Hum Genet 10:513–536
14. Waters AM, Beales PL (2011) Ciliopathies: an expanding disease spectrum. Pediatr Nephrol 26:1039–1056
15. Ridanpaa M, Jain P, McKusick VA, Francomano CA, Kaitila I (2003) The major mutation in the RMRP gene causing CHH among the Amish is the same as that found in most Finnish cases. Am J Med Genet C Semin Med Genet 121C:81–83
16. Morton DH, Bennett MJ, Seargeant LE, Nichter CA, Kelley RI (1991) Glutaric aciduria type I: a common cause of episodic encephalopathy and spastic paralysis in the Amish of Lancaster county, Pennsylvania. Am J Med Genet 41:89–95
17. Morton DH, Strauss KA, Robinson DL, Puffenberger EG, Kelley RI (2002) Diagnosis and treatment of maple syrup disease: a study of 36 patients. Pediatrics 109:999–1008
18. Tarasov KV, Sanna S, Scuteri A et al (2009) COL4A1 is associated with arterial stiffness by genome-wide association scan. Circ Cardiovasc Genet 2:151–158
19. Pollin TI, Damcott CM, Shen H et al (2008) A null mutation in human APOC3 confers a favorable plasma lipid profile and apparent cardioprotection. Science 322:1702–1705
20. Wang Y, O'Connell JR, McArdle PF et al (2009) From the cover: whole-genome association study identifies STK39 as a hypertension susceptibility gene. Proc Natl Acad Sci U S A 106:226–231
21. Cheng YC, Kao WH, Mitchell BD et al (2010) Genetic effects on postprandial variations of inflammatory markers in healthy individuals. Obesity (Silver Spring) 18:1417–1422
22. Daley E, Streeten EA, Sorkin JD et al (2010) Variable bone fragility associated with an Amish COL1A2 variant and a knock-in mouse model. J Bone Miner Res 25:247–261
23. Ma L, Hanson RL, Traurig MT et al (2010) Evaluation of A2BP1 as an obesity gene. Diabetes 59:2837–2845

24. McArdle PF, Parsa A, Chang YP et al (2008) Association of a common nonsynonymous variant in GLUT9 with serum uric acid levels in Old Order Amish. Arthritis Rheum 58:2874–2881
25. Musunuru K, Post WS, Herzog W et al (2010) Association of single nucleotide polymorphisms on chromosome 9p21.3 with platelet reactivity: a potential mechanism for increased vascular disease. Circ Cardiovasc Genet 3:445–453
26. Njajou OT, Blackburn EH, Pawlikowska L et al (2010) A common variant in the telomerase RNA component is associated with short telomere length. PLoS One 5:e13048
27. Parsa A, Chang YP, Kelly RJ et al (2011) Hypertrophy-associated polymorphisms ascertained in a founder cohort applied to heart failure risk and mortality. Clin Transl Sci 4:17–23
28. Roghmann MC, Johnson JK, Stine OC et al (2011) Persistent *Staphylococcus aureus* colonization is not a strongly heritable trait in Amish families. PLoS One 6:e17368
29. Active Amish walk away from obesity genes (2009) Mayo Clin Health Lett 27:4
30. Ahmad Z, Garg A (2011) Lack of cardiovascular disease among Old Order Amish with familial defective apolipoprotein B. Arch Intern Med 171:1039–1040, author reply 1040
31. Arnett FC, Enlow RW, Bias WB (1982) Sacroiliitis in the Old Order Amish. Am J Med Genet 12:333–342
32. Bias WB (1994) Complex segregation analysis of antibodies to thyroid peroxidase in Old Order Amish families. Am J Med Genet 51:86
33. Cheng YC, Kao WH, Mitchell BD et al (2009) Genome-wide association scan identifies variants near Matrix Metalloproteinase (MMP) genes on chromosome 11q21-22 strongly associated with serum MMP-1 levels. Circ Cardiovasc Genet 2:329–337
34. Clayton RJ, Iber FL, Ruebner BH, McKusick VA (1969) Byler disease. Fatal familial intrahepatic cholestasis in an Amish kindred. Am J Dis Child 117:112–124
35. Courtenay MD, Gilbert JR, Jiang L et al (2012) Mitochondrial haplogroup X is associated with successful aging in the Amish. Hum Genet 131(2):201–208
36. Crosby AH, Patel H, Chioza BA et al (2010) Defective mitochondrial mRNA maturation is associated with spastic ataxia. Am J Hum Genet 87:655–660
37. Cross HE, Kennel EE, Lilienfeld AM, McKusick VA (1967) Cancer of the cervix in the Amish. Trans Assoc Am Physicians 80:133–141
38. Cummings AC, Lee SL, McCauley JL et al (2011) A genome-wide linkage screen in the Amish with Parkinson disease points to chromosome 6. Ann Hum Genet 75:351–358
39. Donohoue PA, Guethlein L, Collins MM et al (1995) The HLA-A3, Cw6, B47, DR7 extended haplotypes in salt losing 21-hydroxylase deficiency and in the Old Order Amish: identical class I antigens and class II alleles with at least two crossover sites in the class III region. Tissue Antigens 46:163–172
40. Donohoue PA, Van Dop C, Migeon CJ, McLean RH, Bias WB (1987) Coupling of HLA-A3, Cw6, Bw47, DR7 and a normal CA21HB steroid 21-hydroxylase gene in the Old Order Amish. J Clin Endocrinol Metab 65:980–986
41. Dotson CD, Shaw HL, Mitchell BD, Munger SD, Steinle NI (2010) Variation in the gene TAS2R38 is associated with the eating behavior disinhibition in Old Order Amish women. Appetite 54:93–99
42. Edwards DR, Gilbert JR, Jiang L et al (2011) Successful aging shows linkage to chromosomes 6, 7, and 14 in the Amish. Ann Hum Genet 75:516–528
43. Francomano CA, McKusick VA, Biesecker LG (2003) Medical genetic studies in the Amish: historical perspective. Am J Med Genet C Semin Med Genet 121C:1–4
44. Fuchs T, Gavarini S, Saunders-Pullman R et al (2009) Mutations in the THAP1 gene are responsible for DYT6 primary torsion dystonia. Nat Genet 41:286–288
45. Ghanekar J, Sangrampurkar S, Hulinaykar R, Ahmer T (2009) Ellis-van Creveld syndrome. J Assoc Physicians India 57:532–534
46. Gillum DR, Staffileno BA (2011) An integrative review of the current knowledge of cardiovascular disease and associated risk factors in the Old Order Amish. J Transcult Nurs 22:182–190
47. Hostetler JA (1963) Amish society. Johns Hopkins University Press, Baltimore

48. Houlden H, Schneider SA, Paudel R et al (2010) THAP1 mutations (DYT6) are an additional cause of early-onset dystonia. Neurology 74:846–850
49. Hsu SH, Chan MM, Bias WB (1979) CML target determinant segregating independently of HLA in an Old Order Amish family. Transplant Proc 11:1816–1819
50. Hsu SH, Lambert P, Scribner PP, Hartzman RJ, Robbins F, Bias WB (1983) Primary and secondary MLR characterization of three allelic variants of HLA-Dw7 segregating in a single family. Hum Immunol 6:167–176
51. Huff MO, Li XP, Ginns E, El-Mallakh RS (2010) Effect of ethacrynic acid on the sodium- and potassium-activated adenosine triphosphatase activity and expression in Old Order Amish bipolar individuals. J Affect Disord 123:303–307
52. Jackman C, Horn ND, Molleston JP, Sokol DK (2009) Gene associated with seizures, autism, and hepatomegaly in an Amish girl. Pediatr Neurol 40:310–313
53. Kelly TE, Chase GA, Kaback MM, Kumor K, McKusick VA (1975) Tay-Sachs disease: high gene frequency in a non-Jewish population. Am J Hum Genet 27:287–291
54. Khoury MJ, Cohen BH, Diamond EL, Chase GA, McKusick VA (1987) Inbreeding and pre-reproductive mortality in the Old Order Amish. III. Direct and indirect effects of inbreeding. Am J Epidemiol 125:473–483
55. Khoury MJ, Cohen BH, Diamond EL, Chase GA, McKusick VA (1987) Inbreeding and pre-reproductive mortality in the Old Order Amish. I. Genealogic epidemiology of inbreeding. Am J Epidemiol 125:453–461
56. Khoury MJ, Cohen BH, Newill CA, Bias W, McKusick VA (1987) Inbreeding and prereproductive mortality in the Old Order Amish. II. Genealogic epidemiology of prereproductive mortality. Am J Epidemiol 125:462–472
57. Kwiterovich PO Jr, Bachorik PS, Smith HH et al (1981) Hyperapobetalipoproteinaemia in two families with xanthomas and phytosterolaemia. Lancet 1:466–469
58. Lahiry P, Wang J, Robinson JF et al (2009) A multiplex human syndrome implicates a key role for intestinal cell kinase in development of central nervous, skeletal, and endocrine systems. Am J Hum Genet 84:134–147
59. Lee WJ, Pollin TI, O'Connell JR, Agarwala R, Schaffer AA (2010) PedHunter 2.0 and its usage to characterize the founder structure of the Old Order Amish of Lancaster county. BMC Med Genet 11:68
60. Lie H, Zariwala MA, Helms C et al (2010) Primary ciliary dyskinesia in Amish communities. J Pediatr 156:1023–1025
61. Lohr NJ, Molleston JP, Strauss KA et al (2010) Human ITCH E3 ubiquitin ligase deficiency causes syndromic multisystem autoimmune disease. Am J Hum Genet 86:447–453
62. McKusick VA (2000) Ellis-van Creveld syndrome and the Amish. Nat Genet 24:203–204
63. McKusick VA (1980) The Amish. Endeavour 4:52–57
64. McKusick VA (1973) Genetic studies in American inbred populations with particular reference to the Old Order Amish. Isr J Med Sci 9:1276–1284
65. McKusick VA, Bias WB, Norum RA, Cross HE (1967) Blood groups in two Amish demes. Humangenetik 5:36–41
66. McKusick VA, Egeland JA, Eldridge, R, Krusen DE (1964a). Dwarfism in the Amish I. the Ellis-van Creveld syndrome. Bull Johns Hopkins Hosp 115:306–335
67. McKusick VA, Eldridge R, Hostetler JA, Egeland JA (1964b) Dwarfism in the Amish. Trans Assoc Am Physicians 77:151–168
68. McKusick VA, Hostetler JA, Egeland JA (1964c) Genetic studies of the Amish. Background and potentialities. Bull Johns Hopkins Hosp 115:203–222
69. McKusick VA, Hostetler JA, Egeland JA, Eldridge R (1964d) The distribution of certain genes in the Old Order Amish. Cold Spring Harb Symp Quant Biol 29:99–114
70. Meyers DA, Bias WB, Marsh DG (1982) A genetic study of total IgE levels in the Amish. Hum Hered 32:15–23
71. Meyers DA, Hasstedt SJ, Marsh DG et al (1983) The inheritance of immunoglobulin E: genetic linkage analysis. Am J Med Genet 16:575–581

72. Naj AC, Kao WH, O'Connell JR, Mitchell BD, Silver KD (2009) Sequence variation in IGF1R is associated with differences in insulin levels in nondiabetic Old Order Amish. Diabetes Metab Res Rev 25:773–779
73. Neuhauser G, Wiffler C, Opitz JM (1976) Familial spastic paraplegia with distal muscle wasting in the Old Order Amish; atypical Troyer syndrome or "new" syndrome. Clin Genet 9:315–323
74. Patel H, Cross H, Proukakis C et al (2002) SPG20 is mutated in Troyer syndrome, an hereditary spastic paraplegia. Nat Genet 31:347–348
75. Racette BA, Good LM, Kissel AM, Criswell SR, Perlmutter JS (2009) A population-based study of Parkinsonism in an Amish community. Neuroepidemiology 33:225–230
76. Rampersaud E, Bielak LF, Parsa A et al (2008) The association of coronary artery calcification and carotid artery intima-media thickness with distinct, traditional coronary artery disease risk factors in asymptomatic adults. Am J Epidemiol 168:1016–1023
77. Rampersaud E, Mitchell BD, Pollin TI et al (2008) Physical activity and the association of common FTO gene variants with body mass index and obesity. Arch Intern Med 168: 1791–1797
78. Rider NL, Morton DH, Puffenberger E, Hendrickson CL, Robinson DL, Strauss KA (2009) Immunologic and clinical features of 25 Amish patients with RMRP 70 A–> G cartilage hair hypoplasia. Clin Immunol 131:119–128
79. Sanna S, Busonero F, Maschio A et al (2009) Common variants in the SLCO1B3 locus are associated with bilirubin levels and unconjugated hyperbilirubinemia. Hum Mol Genet 18:2711–2718
80. Sharathkumar A, Hardesty B, Greist A et al (2009) Variability in bleeding phenotype in Amish carriers of haemophilia B with the 31008 C–> T mutation. Haemophilia 15:91–100
81. Shen H, Bielak LF, Ferguson JF, Streeten EA, et al. (2010a) Association of the vitamin D metabolism gene CYP24A1 with coronary artery calcification. Arterioscler Thromb Vasc Biol. 30(12):2648–54
82. Shen H, Damcott CM, Rampersaud E, Pollin TI, et al. (2010b) Familial defective apolipoprotein B-100 and increased low-density lipoprotein cholesterol and coronary artery calcification in the Old Order Amish. Arch Intern Med. 170(20):1850–5
83. Shen H, Pollin TI, Damcott CM, McLenithan JC, Mitchell BD, Shuldiner AR (2009) Glucokinase regulatory protein gene polymorphism affects postprandial lipemic response in a dietary intervention study. Hum Genet 126:567–574
84. Sherman EA, Strauss KA, Tortorelli S et al (2008) Genetic mapping of glutaric aciduria, type 3, to chromosome 7 and identification of mutations in c7orf10. Am J Hum Genet 83:604–609
85. Shuldiner AR, O'Connell JR, Bliden KP et al (2009) Association of cytochrome P450 2C19 genotype with the antiplatelet effect and clinical efficacy of clopidogrel therapy. JAMA 302:849–857
86. Simpson MA, Cross HE, Cross L, Helmuth M, Crosby AH (2009) Lethal cystic kidney disease in Amish neonates associated with homozygous nonsense mutation of NPHP3. Am J Kidney Dis 53:790–795
87. Spiegel R, Shaag A, Edvardson S et al (2009) SLC25A19 mutation as a cause of neuropathy and bilateral striatal necrosis. Ann Neurol 66:419–424
88. Strauss KA, Donnelly P, Wintermark M (2010a) Cerebral haemodynamics in patients with glutaryl-coenzyme A dehydrogenase deficiency. Brain 133:76–92
89. Strauss KA, Morton DH (2003) Branched-chain ketoacyl dehydrogenase deficiency: maple syrup disease. Curr Treat Options Neurol 5:329–341
90. Strauss KA, Wardley B, Robinson D, et al. (2010) Classical maple syrup urine disease and brain development: principles of management and formula design. Mol Genet Metab. 99(4):333–45
91. Sulisalo T, Francomano CA, Sistonen P et al (1994) High-resolution genetic mapping of the cartilage-hair hypoplasia (CHH) gene in Amish and Finnish families. Genomics 20:347–353

92. Trojak JE, Polmar SH, Winkelstein JA et al (1981) Immunologic studies of cartilage-hair hypoplasia in the Amish. Johns Hopkins Med J 148:157–164
93. Westman JA, Ferketich AK, Kauffman RM et al (2010) Low cancer incidence rates in Ohio Amish. Cancer Causes Control 21:69–75
94. Wojciechowski R, Bailey-Wilson JE, Stambolian D (2009) Fine-mapping of candidate region in Amish and Ashkenazi families confirms linkage of refractive error to a QTL on 1p34-p36. Mol Vis 15:1398–1406
95. Wojciechowski R, Stambolian D, Ciner E, Ibay G, Holmes TN, Bailey-Wilson JE (2009) Genomewide linkage scans for ocular refraction and meta-analysis of four populations in the myopia family study. Invest Ophthalmol Vis Sci 50:2024–2032
96. Xiao J, Zhao Y, Bastian RW et al (2010) Novel THAP1 sequence variants in primary dystonia. Neurology 74:229–238
97. Xin B, Jones S, Puffenberger EG et al (2011) Homozygous mutation in SAMHD1 gene causes cerebral vasculopathy and early onset stroke. Proc Natl Acad Sci USA 108:5372–5377
98. Xin B, Puffenberger EG, Turben S, Tan H, Zhou A, Wang H (2010) Homozygous frameshift mutation in TMCO1 causes a syndrome with craniofacial dysmorphism, skeletal anomalies, and mental retardation. Proc Natl Acad Sci USA 107:258–263
99. Yang S, Wang K, Gregory B et al (2009) Genomic landscape of a three-generation pedigree segregating affective disorder. PLoS One 4:e4474
100. Eaton JW and Mayer AJ (1953) The social biology of very high fertility among the Hutterites. The demography of a unique population
101. Juberg RC (1966) Selection in the ABO, Rhesus, and MNSs Blood group polymorphisms in an Amish isolate of northern Indiana. Ph.D. Thesis, University of Michigan
102. Murdoch JL and Walker BA (1969) Ellis-van Creveld syndrome. The Clinical Delineation of Birth Defects IV. Skeletal Dysplasias. Birth Defects Original Article Series V, No. 4 pp. 279–288
103. Nevanlinna, HR (1972) The Finnish population structure: A genetic and genealogic study. Hereditas 71:195–236
104. Nevanlinna HR (1977) Personal communication to VA McKusick, cited in 11, p. 280
105. Ruiz-Perez VL, Ide SE, Strom TM et al (2000) Mutations in a new gene in Ellis-van Creveld syndrome and Weyers acrodental dysostosis. Nat Genet. 24(3):283–286
106. Ruiz-Perez VL, Tompson SW, Blair HJ et al (2003) Mutations in two nonhomologous genes in a head-to-head configuration cause Ellis-van Creveld syndrome. Am J Hum Genet. 72(3):728–732
107. Ruiz-Perez VL, Blair HJ, Rodriguez-Andres ME et al (2007) Evc is a positive mediator of Ihh-regulated bone growth that localises at the base of chondrocyte cilia. Development 134(16):2903–2912
108. Ridanpää M, van Eenennaam H, Pelin K et al (2001) Mutations in the RNA component of RNase MRP cause a pleiotropic human disease, cartilage-hair hypoplasia. Cell 104(2):195–203
109. Tompson SW, Ruiz-Perez VL, Blair HJ, et al (2007) Sequencing EVC and EVC2 identifies mutations in two-thirds of Ellis-van Creveld syndrome patients. Hum Genet. 120:663–670

Chapter 14
Dr. Victor A. McKusick and the Genetics of Dwarfism

Clair A. Francomano and David L. Rimoin†

When Dr. Victor McKusick was awarded the Lasker prize in 1997, he wrote a commentary on his 50 years of observations concerning intestinal polyposis, Marfan syndrome, and achondroplasia [12]. He notes in that paper that achondroplasia came to his interest early in his studies of medical genetics, in keeping with his interest with the heritable disorders of connective tissue. His studies of the Old Order Amish began in 1963, in part, because of a report that "achondroplastic dwarfism" was unusually frequent in that population. Because achondroplasia is an autosomal dominant disorder with more than 80% of cases arising as a result of new mutations, McKusick postulated that the skeletal dysplasia in the Amish was another condition inherited as an autosomal recessive trait.

McKusick's studies eventually led to the identification of two distinct phenotypes in the Amish community. These were the Ellis-van Creveld syndrome, a previously described, rare, skeletal dysplasia with polydactyly [6, 13], and cartilage-hair hypoplasia, a newly recognized skeletal dysplasia associated with a T cell immune deficiency and fine hair lacking a central pigment core [7].

As a result of his studies in the Amish community, in 1965, McKusick became involved with the nascent Little People of America (LPA), a support group for persons of short stature and their families. He recognized that among the membership of the LPA were multiple distinct clinical entities that had not, at the time, been delineated [10]. At that time, the nosology of short stature was largely defined by short-limbed dwarfism, as in achondroplasia, and short-trunked dwarfism, as

C.A. Francomano, M.D. (✉)
Harvey Institute for Human Genetics, Greater Baltimore Medical Center, 6701 North Charles Street, Baltimore, MD 21204, USA
e-mail: cfrancomano@gbmc.org

D.L. Rimoin, M.D., Ph.D.
Medical Genetics Institute, Cedars-Sinai Medical Center, 8700 Beverly Blvd., Los Angeles, CA 90048, USA
e-mail: David.Rimoin@cshs.org

K.R. Dronamraju and C.A. Francomano (eds.), *Victor McKusick and the History of Medical Genetics*, DOI 10.1007/978-1-4614-1677-7_14,

typified by Morquio syndrome. McKusick and his fellows in medical genetics, including Charles Scott, David Rimoin, Judith Hall, Roswell Eldridge, and many others, started delineating the clinical features that set distinct dwarfism phenotypes apart. Achondroplasia, pseudoachondroplasia, hypochondroplasia, and diastrophic dysplasia among many others were explored and described in detail during the early 1970s [5, 16, 17, 18, 22, 26, 27], and the various mucopolysaccharidoses were similarly dissected during that time [9, 23]. Two skeletal dysplasias, osteogenesis imperfect and Hurler syndrome, were among the original disorders described in McKusick's first edition of Heritable Disorders of Connective Tissue.

To take a slight step backward, even before McKusick become involved with the Amish, in 1964, David Rimoin was a fellow with McKusick. He saw a woman who had growth hormone deficiency with sexual maturity. Conventional wisdom at the time held that pituitary dwarfism involved both growth hormone deficiency and hypogonadism; this observation established that it is indeed possible to have growth hormone deficiency without hypogonadism [20].

Shortly thereafter, Rimoin came across a volume of National Geographic describing the pygmies of Central Africa. He recognized that they appeared quite similar to his recent patient with growth hormone deficiency. Within weeks, he was traveling to Africa to investigate the endocrine aspects of the pygmies [15, 19, 21]. McKusick made arrangements with Luca Cavalli Sforza to meet Rimoin in the Central African Republic and facilitate the investigations which showed that the pygmies had normal levels of growth hormone but were resistant to its actions [14]. Another finding made through these studies was the recognition that growth hormone and prolactin were distinct proteins and that one could be deficient in growth hormone without being deficient in prolactin [20]. They learned this through a woman with isolated growth hormone deficiency who was able to actively nurse her child; the presence of lactation with growth hormone deficiency spoke to the distinct nature of prolactin and growth hormone as molecules.

After the first meeting in 1965, attendance at the annual meetings of the Little People of America became a regular feature of the yearly calendar for McKusick and the fellows in the Medical Genetics Program. Attendees at the meeting would sign up for consultations with geneticists, orthopedists, neurologists, and neurosurgeons who provided these consultations at no expense. Over lunch during the convention, interesting cases would be presented and discussed. This practice continues to this day.

For many years, the Moore Clinic (the genetics clinic at Johns Hopkins founded by McKusick in the mid-1960s) hosted a Short Stature Symposium, in which persons of short stature would convene for clinical testing, workshops, and camaraderie. McKusick started the March of Dimes Birth Defects Symposia on the Clinical Delineation of Birth Defects in 1968, which he hosted in Baltimore for 5 years. This first meeting resulted in the publication of one of the first large volumes on the skeletal dysplasias [2]. In 1979, the national convention of the Little People of America was held in Lancaster County, Pennsylvania, the center of the Amish community. McKusick hosted a Short Stature Symposium at Hopkins just prior to the convention.

During the LPA convention, Sara Fisher, his liaison to the Amish for many years, guided tours for the membership of the LPA to the various sites around Lancaster County. McKusick was awarded a lifetime membership in the Little People of America for his contributions to the care of persons with short stature.

Mrs. Kay Smith became the coordinator for short stature very early in the course of McKusick's studies of skeletal dysplasias. She attended all the Little People of America meetings and knew all of the patients who came to Johns Hopkins by name. Moreover, she started fund-raising for research and education into causes of short stature. Her handwritten letters and personal attention to every donation that came in eventually led to the establishment of the Greenberg Center for Skeletal Dysplasias at Johns Hopkins. Kay Smith's role in the genesis of the study of skeletal dysplasias at Johns Hopkins was pivotal. The Greenberg Center for Skeletal Dysplasias is now led by Dr. Julie Hoover-Fong, and for the past two decades, Kathleen "Dee" Miller has served as the coordinator.

As McKusick became involved with more and more diagnoses causing short stature, and describing the clinical phenotypes, radiographic features, and pathology of the various entities he came to recognize, his colleagues at Johns Hopkins were pulled into his orbit and began to investigate from the perspective of their own disciplines. Dr. John Tyson, in obstetrics and gynecology, wrote about the reproductive aspects of achondroplasia and other skeletal dysplasias [24, 25]. Dr. Steven Kopits, an orthopedist at Johns Hopkins, eventually devoted his career to caring for patients with skeletal dysplasias [4]. Dr. John Dorst of pediatric radiology became one of a handful of radiologists worldwide with expertise in the skeletal dysplasias. Dr. Benjamin Carson, in the early days of his illustrious career as a pediatric neurosurgeon, pioneered surgery for foramen magnum stenosis in young children with achondroplasia.

In his early writings about hypochondroplasia and achondroplasia, McKusick proposed that they were likely allelic disorders based on the similarities in clinical and radiographic findings [11]. Four decades later, the gene for achondroplasia was reported by John Wasmuth and his colleagues to be FGFR3, and shortly thereafter, Gary Bellus, working in the Francomano laboratory, reported that hypochondroplasia was also caused by FGFR3 mutations [1], establishing the veracity of McKusick's clinical hypothesis.

In 1974, McKusick, with Dr. Stephen Koptis and others, reported on the "pathodynamics and management" of pseudoachondroplastic dwarfism [5]. Somatic mosaicism in pseudoachondroplasia was later reported by Dr. Judith Hall, McKusick, and others in a New York family of Italian descent [3].

At the very first conference on the nosology of human skeletal dysplasias, McKusick was the United States representative in Europe, meeting with Maroteaux and Jurgen Spranger, among others. From that meeting evolved the International Skeletal Dysplasia Society, a nomenclature subcommittee of which meets every 4 years to review and refine the nosology.

Victor McKusick has been called the "father of medical genetics." Perhaps we should now revise that designation to the "grandfather of medical genetics."

Certainly, this is true if we consider the field of skeletal dysplasias. His contributions to this arena truly cannot be overstated. His trainees, and the trainees of those trainees, have led the way in describing the clinical features of the skeletal dysplasias and their genetics over the past 60 years.

References

1. Bellus GA, McIntosh I, Smith EA et al (1995) A recurrent mutation in the tyrosine kinase domain of fibroblast growth factor receptor 3 causes hypochondroplasia. Nat Genet 10:357–359
2. Bergsma D, McKusick VA, Dorst JP, Scott CI (eds) (1969) The first conference on the clinical delineation of birth defects. Part IV. Skeletal dysplasias. Birth Defects Orig Art Ser, Roswell Eldridgees 5(4):1–396
3. Hall JG, Dorst JP, Rotta J, McKusick VA (1987) Gonadal mosaicism in pseudoachondroplasia. Am J Med Genet 28:143–151
4. James AE Jr, Dorst JP, Mathews ES, McKusick VA (1972) Hydrocephalus in achondroplasia studied by cisternography. Pediatrics 49:46–49
5. Kopits SE, Lindstrom JA, McKusick VA (1974) Pseudoachondroplastic dysplasia: pathodynamics and management. Birth Defects Orig Artic Ser 10:341–352
6. McKusick VA, Egeland JA, Eldridge R, Krusen DE (1964) Dwarfism in the Amish I. The Ellis-Van Creveld syndrome. Bull Johns Hopkins Hosp 115:306–336
7. McKusick VA, Eldridge R, Hostetler JA, Ruangwit U, Egeland JA (1965) Dwarfism in the Amish. II. Cartilage-hair hypoplasia. Bull Johns Hopkins Hosp 116:285–326
8. McKusick VA, Mahloudji M, Abbott MH, Lindenberg R, Kepas D (1967) Seckel's bird-headed dwarfism. N Engl J Med 277:279–286
9. McKusick VA (1969) The nosology of the mucopolysaccharidoses. Am J Med 47:730–747
10. McKusick VA (1969) On lumpers and splitters, or the nosology of genetic disease. Perspect Biol Med 12:298–312
11. McKusick VA, Kelly TE, Dorst JP (1973) Observations suggesting allelism of the achondroplasia and hypochondroplasia genes. J Med Genet 10:11–16
12. McKusick VA (1997) Observations over 50 years concerning intestinal polyposis, Marfan syndrome and achondroplasia. Nat Med 3:1065–1068
13. McKusick VA (2000) Ellis-van Creveld syndrome and the Amish. Nat Genet 24:203–204
14. Merimee TJ, Rimoin DL, Cavalli-Sforza LC, Rabinowitz D, McKusick VA (1968) Metabolic effects of human growth hormone in the African pygmy. Lancet 2:194–195
15. Merimee TJ, Rimoin DL, Rabinowitz D, Cavalli-Sforza LL, McKusick VA (1968) Metabolic studies in the African pygmy. Trans Assoc Am Physicians 81:221–230
16. Miller JD, McKusick VA, Malvaux P, Temtamy S, Salinas C (1975) The 3-M syndrome: a heritable low birthweight dwarfism. Birth Defects Orig Artic Ser 11:39–47
17. Murdoch JL, Walker BA, Hall JG, Abbey H, Smith KK, McKusick VA (1970) Achondroplasia – a genetic and statistical survey. Ann Hum Genet 33:227–244
18. Rimoin DL, Fletcher BD, McKusick VA (1968) Spondylocostal dysplasia. A dominantly inherited form of short-trunked dwarfism. Am J Med 45:948–953
19. Rimoin DL, Merimee TJ, Rabinowitz D, Cavalli-Sforza LL, McKusick VA (1969) Peripheral subresponsiveness to human growth hormone in the African pygmies. N Engl J Med 281:1383–1388
20. Rimoin DL, Holzman GB, Merimee TJ et al (1968) Lactation in the absence of human growth hormone. J Clin Endocrinol Metab 28:1183–1188
21. Rimoin DL, Merimee TJ, Rabinowitz D, McKusick VA (1968) Genetic aspects of clinical endocrinology. Recent Prog Horm Res 24:365–437

22. Siggers CD, Rimoin DL, Dorst JP et al (1974) The Kniest syndrome. Birth Defects Orig Artic Ser 10:193–208
23. Spranger JW, Koch F, McKusick VA, Natzschka J, Wiedemann HR, Zellweger H (1970) Mucopolysaccharidosis IV (Maroteaux-Lamy's disease). Helv Paediatr Acta 25:337–362
24. Tyson JE, Barnes AC, McKusick VA, Scott CI, Jones GS (1970) Obstetric and gynecologic considerations of dwarfism. Am J Obstet Gynecol 108:688–704
25. Tyson JE, Barnes AC, Merimee TJ, McKusick VA (1970) Isolated growth hormone deficiency: studies in pregnancy. J Clin Endocrinol Metab 31:147–152
26. Walker BA, Murdoch JL, McKusick VA, Langer LO, Beals RK (1971) Hypochondroplasia. Am J Dis Child 122:95–104
27. Walker BA, Scott CI, Hall JG, Murdoch JL, McKusick VA (1972) Diastrophic dwarfism. Medicine 51:41–59

Chapter 15
Clinical Genomicist in the Future of Medical Practice

Edison T. Liu

Medical subspecialties commonly arise from either primarily academic origins or because of practical need. A subspecialty often coalesces around academic thought leaders or around the practical need for division of labor in day-to-day clinical practice. In so many cases, from gastroenterology to invasive cardiology and to neuroradiology, it was the availability of innovative and powerful technologies that spawned new clinical subspecialties. In the last fifteen years, human genetics has been transformed by the emergence of genomics as a scientific discipline fuelled by the new sequencing technologies and associated computational capabilities. We believe, in the wake of this genomics movement a new subspecialty, clinical genomics, will emerge that will become an essential part of medical practice. Because of the complexity of the genome and its relationship to any phenotype, a physician expert will have to act as the interpreter of the genome to medical practitioners. In this aspect, clinical genomics will resemble diagnostic radiology. In radiology, complex signals are computationally converted into recognizable images interpreted by a physician-expert, the radiologist. Similarly, genomics is based on complex sequencing technologies whose data output requires computational transformation so that a human expert can interpret the biological meaning of the results.

Victor McKusick was one of the pioneering academic thought leaders who codified human genetics as a medical science and a medical specialty. He organized genetic diagnosis into a reproducible and useful knowledge base; he provided mechanistic understanding of genetic disorders, and he fostered the integration of genetic diagnostic tools from cytogenetics to clinical sequencing into standard practice. His scientific work, his mentoring of hundreds of intellectual disciples, his organizational skills, and his force of personality helped fuel the development of clinical genetics as an essential clinical discipline. Long before "genomics" was

E.T. Liu, M.D. (✉)
The Jackson Laboratory, Bar Harbor, ME 04609, USA
e-mail: liue@gis.a-star.edu.sg; edison.liu@jax.org

K.R. Dronamraju and C.A. Francomano (eds.), *Victor McKusick and the History of Medical Genetics*, DOI 10.1007/978-1-4614-1677-7_15,

coined, Victor McKusick had initiated MIM, a database of Mendelian diseases in man. This database is critical piece for translating raw sequencing data into meaningful clinical genomic information. He had the foresight to integrate human genetic mutations with clinical phenotypes, molecular biology, and disease etiology. Its current format, OMIM (Online Mendelian Inheritance in Man), is one of the foundational tools in genomic medicine used to prioritize disease genes and to clinically annotate disease mutations.

In the last 15 years, human genetics has been transformed by the emergence of genomics as a scientific discipline fuelled by the new sequencing technologies and associated computational capabilities. The fundamentals of genomics are firmly grounded on genetic principles, but the expanded scope and complexity of these approaches and the scale of knowledge generated have begun to have a profound impact on clinical genetics. We believe, in the wake of this genomics movement, a new subspecialty, clinical genomics, will emerge that will become an essential part of medical practice. What might clinical genomics be, what is the clinical need, how might it coalesce as a clinical discipline, and what benefits it might bring will be explored in this commentary. Our basic premise is that clinical genomics will emerge as a practical subspecialty providing expertise in interpreting complex genomic and genetic information for referring physicians.

While genetics is primarily the study of individual genes and their biological effects, genomics is the study of all genes and genetic elements in the genome and how they may function individually or collectively to have an impact on phenotype. Genetics was a discipline based on inference, whereas genomics, because of the precision and comprehensiveness of sequencing technologies, can provide exact mutational information. Theoretically, what was inferred can now be directly sequenced. The challenges now are how to interpret the sequence mutations as to whether they indeed perturb function and how to assess the impact of multiple gene-gene and gene-environment interactions and complex regulatory control of transcription. The difference, therefore, is not one involving fundamentals but one of functional attribution, scale, and complexity.

The clinical applications of genomic sciences and sequence technologies have made a difference. Genomic-based expression profiling has now a place in cancer diagnostics to direct therapeutic choice [1, 2] especially in breast cancer. Array comparative genomic hybridization (array CGH) is the first-line diagnostic for childhood mental retardation and other congenital disorders caused by copy number mutations [3, 4]. Selective sequencing is identifying rare or private coding mutations in familial diseases and discovering targetable functional mutations in cancer oncogenes such as BRAF1 in melanoma [5] and EML4-ALK [6] and EGFR mutations in lung cancer [7]. Prenatal diagnostics of serum DNA using the statistical power of bulk sequencing can discern trisomic states in circulating fetal DNA [8]. The collection of genome-wide association (GWAS) studies suggests tens to hundreds of common risk alleles contributing to diseases such as type I and II diabetes, autoimmune disorders, cancer, and cardiometabolic maladies [9, 10]. The ultimate convergence will be in whole genome sequencing where nucleotide changes, copy number variations, and structural differences can be ascertained using a single technology and in one diagnostic action. We posit that all of genetic diagnostics will therefore converge through this single technology. The data output, however, will be highly complex and nuanced. How to interpret disease risk with complex gene-gene interactions, varying regulatory cassettes, and a wide range of clinical impact of any risk allele is a daunting prospect and highly unlikely to be completely automated using computational risk models. Instead, we believe that a physician expert will have to act as the interpreter of the genome to medical practitioners. In this aspect, clinical genomics will resemble diagnostic radiology.

The historical development of radiology as a science and medical discipline is a surprisingly apt model for clinical genomics. Radiology is fundamentally based on technologies arising from physics but revolutionized medicine because it allowed physicians to peer into the anatomical details of the human body noninvasively. In all contemporary radiological technologies, complex signals are computationally converted into recognizable images interpreted by a physician expert, the radiologist. The complexity of the pixel images has thwarted most attempts at machine-automated interpretation, thus still requiring an expert interface. Similarly, genomics is based on complex sequencing technologies whose data output requires computational transformation so that a human expert can interpret the biological meaning of the results. The complexity of the data is such that a final medical interpretation will still need further integration by the genetic expert.

What might we expect then? Subspecialty clinics will refer specific patients for whole genome sequencing to resolve a diagnostic or risk determination issue. These patient genomes will be sequenced in GLP quality laboratories, and data will be transmitted to analytics centers. This information will undergo data transformation into a format that can be presented on a computer console and then read by an expert clinical genomicist. Available to this expert will be the catalog of appropriate information that can be used for functional and clinical annotation. The clinical genomicist will then integrate patient information with this complex genomic data to arrive at a probabilistic interpretation and recommendation. This genetic interpretation can be discussed with the referring physician so that his/her clinical domain expertise can be incorporated into the therapeutic recommendation. A final written/dictated report can then be part of the medical record. The core of this subspecialty function will be the expert genomicist and the computational analytics that will hopefully be, in large part, automated. This computation-domain expert interface is already commonplace in medicine: sophisticated computational algorithms render CT or MRI digital information into images that mainly a trained radiologist can interpret. Unfortunately, unlike radiology, the state of the computational tools in whole genome sequencing is not yet stable and is insufficiently mature to be used as a routine clinical tool. But the development of such clinically relevant analytic platforms will not only satisfy a clinical need but will also drive a new industry within medical diagnostics that can generate economic activity.

What are some of the unique aspects of this potentially new subspecialty? First, the genome sequence of an individual does not change unless one is dealing with the genome of a cancer. Therefore, theoretically sequencing a patient once is sufficient for most future susceptibility diagnostics in his/her lifetime. The limiting factor today however is the uncertainty of the importance of most sequence variations in the germ line and how they combine to provide a risk profile. Over time, however, such sequence-trait associations will become more clear as greater information is available. Therefore, the clinical genomicist will be practicing a form of archival medicine where the fixed sequence of a patient can be mined at intervals of time as the medical literature becomes more rich with genomic associations. Second, complex microbial diagnostics may very well be another important function for the clinical genomicist, therefore broadening the scope and the training requirements of the discipline. The complex microbial communities in the gut appear to be involved in diabetes, obesity, inflammatory bowel disease, and behavior [11, 12]. Given the difficulty in the isolation and identification of components within bacterial communities, sequencing has been the only technology that can both assess the diversity and quantify the constituents with some precision. Therefore, the serial metagenomic analyses of stool samples may give a nuanced readout of general health.

Should departments or divisions of clinical genetics be taking the lead in advancing clinical genomics as we have outlined? The answer should unequivocally be yes. Scientifically, this would be most appropriate since the fundamentals of human genetics and morbid genetics remain pertinent in the genomic era. The principles of transmission, segregation, and mutation would be the same but can be resolved with greater detail using genomic tools. Professionally, this would also be important for the vibrancy of the clinical genetics subspecialty to be engaged in such an expansive and impactful new discipline as genomics.

What does this mean for specialty training in clinical genetics? Since the genomic analysis is complex as are the technologies, we anticipate that training in the computational platforms supporting genetic sequence analytics will become part of the curriculum. Whether there should be further subspecialization into clinical genetics and clinical genomics like that between anatomic and clinical pathology will be determined by demand and utility. However, a key requirement will be some standardization of the analytical platform akin to the instrumentation and algorithm standards in radiologic image processing. This standardization goes beyond establishing a GLP sequencing facility in that computational algorithms will need to be fixed and validated. We anticipate that once standard genome analytics have been codified, then formalization of the training process will follow.

The emergence of clinical genomics as a subspecialty is inevitable given the power of the technologies and the clinical utility of the information. We believe that the clinical genetics community should be prepared to lead in its rise and shape its development.

Acknowledgment The author thanks Dr. Pauline Ng for her contributions to this chapter.

Victor McKusick, Gail Martin, Neal Copeland, Harold Varmus, Lee Hood, Peter Agre, Edward Rubin, Peter Doherty (photo credit to "The Jackson Laboratory")

Bar Harbor Course 2004 course group (photo credit to "The Jackson Laboratory")

An early Bar Harbor Course in Medical Genetics, Dr. Victor McKusick is standing on the right side in the front. Courtesy of The Jackson Laboratory, Bar Harbor, ME

References

1. Mook S, Schmidt MK, Viale G, Pruneri G, Eekhout I, Floore A, Glas AM, Bogaerts J, Cardoso F, Piccart-Gebhart MJ, Rutgers ET, Van't Veer LJ (2009) TRANSBIG consortium. The 70-gene prognosis-signature predicts disease outcome in breast cancer patients with 1-3 positive lymph nodes in an independent validation study. Breast Cancer Res Treat 116(2):295–302
2. Kim C, Paik S (2010) Gene-expression-based prognostic assays for breast cancer. Nat Rev Clin Oncol 7(6):340–347
3. Boone PM, Bacino CA, Shaw CA, Eng PA, Hixson PM, Pursley AN, Kang SH, Yang Y, Wiszniewska J, Nowakowska BA, del Gaudio D, Xia Z, Simpson-Patel G, Immken LL, Gibson JB, Tsai AC, Bowers JA, Reimschisel TE, Schaaf CP, Potocki L, Scaglia F, Gambin T, Sykulski M, Bartnik M, Derwinska K, Wisniowiecka-Kowalnik B, Lalani SR, Probst FJ, Bi W, Beaudet AL, Patel A, Lupski JR, Cheung SW, Stankiewicz P (2010) Detection of clinically relevant exonic copy-number changes by array CGH. Hum Mutat 31(12):1326–1342
4. Pinto D, Pagnamenta AT, Klei L, Anney R, Merico D et al (2010) Functional impact of global rare copy number variation in autism spectrum disorders. Nature 466(7304):368–372
5. Poulikakos PI, Rosen N (2011) Mutant BRAF melanomas – dependence and resistance. Cancer Cell 19(1):11–15
6. Soda M, Choi YL, Enomoto M, Takada S, Yamashita Y, Ishikawa S, Fujiwara S, Watanabe H, Kurashina K, Hatanaka H, Bando M, Ohno S, Ishikawa Y, Aburatani H, Niki T, Sohara Y, Sugiyama Y, Mano H (2007) Identification of the transforming EML4-ALK fusion gene in non-small-cell lung cancer. Nature 448(7153):561–566

7. Pao W, Chmielecki J (2010) Rational, biologically based treatment of EGFR-mutant non-small-cell lung cancer. Nat Rev Cancer 10(11):760–774
8. Dennis Lo Y, Chiu RW (2011) Plasma nucleic acid analysis by massively parallel sequencing: pathological insights and diagnostic implications. J Pathol 225(3):318–323
9. McCarthy MI (2010) Genomics, type 2 diabetes, and obesity. N Engl J Med 363(24):2339–2350
10. Pharoah PD, Antoniou AC, Easton DF, Ponder BA (2008) Polygenes, risk prediction, and targeted prevention of breast cancer. N Engl J Med 358(26):2796–2803
11. Kau AL, Ahern PP, Griffin NW, Goodman AL, Gordon JI (2011) Human nutrition, the gut microbiome and the immune system. Nature 474(7351):327–336
12. Heijtz RD, Wang S, Anuar F, Qian Y, Björkholm B, Samuelsson A, Hibberd ML, Forssberg H, Pettersson S (2011) Normal gut microbiota modulates brain development and behavior. Proc Natl Acad Sci USA 108(7):3047–3052

Chapter 16
Victor McKusick and the History of Medical Genetics

Peter S. Harper

> *History has always interested me greatly. In a sense, genetics is history. The history of our species is written in the human genome. [5]*

Introduction

Victor McKusick's contributions to human gene mapping, to the delineation and classification of genetic disorders, and to the building of medical genetics as a clinical and academic discipline are well recognised and well documented, both by his own writings and by those of others. His role as a historian of science and medicine through his historical approaches to medical genetics is less well recognised, however, though his historical opening chapter in Emery and Rimoin's textbook *Principles and Practice of Medical Genetics* is a lasting reminder of this to the medical genetics community. In some ways, McKusick has been regarded as embodying the history of medical genetics through being himself, rather than through his specific contributions.

Here I try to summarise Victor McKusick's main historical contributions to the field and in particular to make the wider genetics community aware of some of his less well known historical papers. I do not attempt to make a critical or analytical approach to his work; this will be better and more objectively done by historians who can stand back from the field.

P.S. Harper (✉)
Institute of Medical Genetics, Cardiff University School of Medicine,
Heath Park, Cardiff, CF14 4XN, UK
e-mail: HarperPS@cardiff.ac.uk

K.R. Dronamraju and C.A. Francomano (eds.), *Victor McKusick and the History of Medical Genetics*, DOI 10.1007/978-1-4614-1677-7_16,

Victor McKusick as a Historian

As evidenced by the quotation at the start of this chapter, Victor McKusick's interest in the historical aspects of medicine was a lifelong one. Nor was it confined to the field of medical genetics; in his early career as a cardiologist, his book (1958) on cardiovascular sound [29] had an extensive historical section, and McKusick notes in relation to this:

> Preparation of the first chapter, on the history of the field, gave me much pleasure. I have insisted that the best way to get 'on top' of a field is to know how it got where it is now from where it was at the beginning. In each of the fields in which I have worked I have explored its history [1].

This historical approach pervades most of his work and makes it difficult to divide his publications into those that are strictly 'historical' and those that focus on current work, but in Table 16.1, an attempt is made to summarise the broad categories of contribution to the history of medical genetics that McKusick has made and which are discussed more fully in later sections of this chapter.

The Development of Medical Genetics

Over his 60-year career, Victor McKusick was well placed to observe and chronicle the development and progress of the field of medical genetics, from its origins as a defined area of medicine in the late 1950s to its emergence and flowering as a mature medical specialty towards the end of the twentieth century. He did so in a series of general reviews (see Table 16.2), some based on presidential lectures or awards, others written for more general readership in journals such as Journal of the American Medical Association (JAMA).

As a clinical rather than an experimental scientist, McKusick was particularly concerned with how the successive discoveries and new technologies in genetics might be applied to medical genetics practice and in wider medicine and also how the new speciality of medical genetics might best be organised. Although he was thinking here primarily in an American context, the large number of fellows training with him from many countries across the world has given these patterns of organisation wide international implementation.

Table 16.1 Victor McKusick. Historical contributions

General reviews on the progress of medical genetics
Historical reviews on specific topics in medical genetics
Biographical articles on specific individuals
Photographs
Records archive
Personal articles, obituaries and interviews

Table 16.2 The progress of medical genetics, as documented by Victor McKusick

1975	The growth and development of human genetics as a clinical discipline McKusick VA. Am J Hum Genet. 1975 May;27(3):261–273
1981	The last twenty years: an overview of advances in medical genetics McKusick VA. Prog Clin Biol Res. 1981;45:127–144
1989	Forty years of medical genetics McKusick VA. JAMA. 1989 Jun 2;261(21):3155–3158
1989	Historical perspectives: the understanding and management of genetic disorders McKusick VA. Maryland Med J. 1989 Nov;38(11):901–908
1991	Advances in medical genetics in the past 30 years McKusick VA. Res Publ Assoc Res Nerv Ment Dis. 1991;69:1–18
1992	Human Genetics: The Last 35 Years, the present, and the future McKusick VA. Am J Hum Genet. 1992 Apr;50(4):663–670
1993	Medical genetics. A 40-year perspective on the evolution of a medical specialty from a basic science McKusick VA. JAMA. 1993 Nov 17;270(19):2351–2356
1997	1997 Albert Lasker Award for Special Achievement in Medical Science. Observations over 50 years concerning intestinal polyposis, Marfan syndrome and achondroplasia McKusick VA. Nat Med. 1997 Oct;3(10):1065–1068
2006	A 60-Year Tale of spots, maps, and genes McKusick VA. Annu Rev Genomics Hum Genet. 2006;7:1–27
2002 and 2007	History of Medical Genetics (VA McKusick. Chapter 1 pp 3–32 in Emery and Rimoin's Principles and Practice of Medical Genetics; Rimoin et al. 2007)

McKusick's presidential address to the 1991 8th International Congress of Human Genetics, *Human Genetics: The Last 35 Years, the Present, and the Future* [2], gives a striking picture of the scientific advances in each of the 5-year periods marked by the international congresses; the paper is enhanced by the tables and graphs that are reproduced from the lecture itself. Likewise, his cautionary concluding remarks:

> As geneticists we have a responsibility to avoid unfounded conclusions and overblown interpretations and to inculcate profound respect for the genetic variability that is the strength of the species and indeed of the individual.

echo Lionel Penrose's comparable warning 40 years earlier as president of the same congress in 1966:

> At the moment we are only scratching the surface of this great science and our knowledge of human genes and their action is still so slight that it is presumptuous and foolish to lay down positive principles for human breeding. Rather each person can marvel at the prodigious diversity of the hereditary characters in man and respect those who differ from him genetically [3].

McKusick could well have written a full book on the history of medical genetics, but he did not; perhaps he felt too close to the topic and was reluctant to bring himself too much into the picture. The lack of such a book was one of the main reasons for attempting the task myself [4]; it was a privilege for me to have his detailed comments on the draft, full of insight even though he was gravely ill at the time. However, his 50-page chapter, *History of Medical Genetics*, which forms the opening chapter to the definitive textbook on the field, *Emery and Rimoin's Principles and Practice of*

Medical Genetics, forms a valuable substitute for a book, though highly condensed and summarised. It first appears in the third edition of the book, being updated for the 2002 and 2007 editions [44]. It is illustrated largely by McKusick's own photographs, themselves a valuable historical record (see below).

Specific Topics in Medical Genetics

Victor McKusick's combination of an exceptionally long active career (over 60 years), a remarkable memory and a wide-ranging interest over medicine generally gave him special opportunities for following particular themes as they developed over a period of decades, often revisiting them in the context of a particular new discovery or unusual family. His position at Johns Hopkins Hospital provided a vantage point for long continued observations and syntheses of this nature, as he acknowledged in one of his wider reviews [5].

> I have been blessed by a marvellous variety of exciting topics on which to work and an institution of great depth and breadth in which to work. I have been at Johns Hopkins uninterruptedly for more than 46 years and a member of the Faculty continuously since 1947. I have not even taken a sabbatical. Such a sessile career is unusual in academic medicine, which, as a rule, leads to a nomadic existence. The sessile career has been possible because Johns Hopkins is, figuratively and literally, at a crossroads. It is located on the trade routes, let us say, and it has been possible for me to pitch my tent beside the road and find out what goes on in far off Cathay without ever travelling there myself.

It is thus not surprising that some of his more specific historical overviews are 'Johns Hopkins-orientated', for example, his Karl Landsteiner Award article on blood group genetics, prompted by the Nobel Prize given to Hopkins scientist Peter Agre. Likewise his account of the Birth Defects Conferences [6] relates primarily to the series of clinical conferences and their published proceedings held annually at Johns Hopkins, supported by March of Dimes, which provided firm foundations for the new field of dysmorphology.

Especially valuable and in some ways more definitive are McKusick's historical accounts of specific fields of genetics in which he was closely involved himself (see Table 16.3). Although in no sense an experimental scientist, his ability to synthesise and to correlate clinical and scientific data and ideas, along with his remarkable foresight into future trends and developments, meant that his influence in some of these areas was profound.

Human Gene Mapping

Of all McKusick's scientific interests, the field of human gene mapping was undoubtedly closest to his heart, and he was able to be a key part of the 'gene mapping community' from its beginnings in the 1950s to its transformation in the 1990s into the human genome project. Although he and his unit made specific early contributions (e.g. the first assignment of a Mendelian trait to a specific autosome [7]), his

Table 16.3 Specific topics in medical genetics. Victor McKusick's historical contributions

Haematology	Karl Landsteiner award. From Karl Landsteiner to Peter Agre: 100 years in the history of blood group genetics
	McKusick VA. Transfusion. 2004 Sep;44(9):1370–1376
Dermatology	Genetics and dermatology or if I were to rewrite Cockayne's *Inherited Abnormalities of the Skin*
	McKusick VA. 1973. J Invest Dermatol. 60:343–359
Haemophilia	The royal hemophilia
	McKusick VA. Sci Am. 1965 Aug; 213:88–95
Brachydactyly	Farabee's brachydactylous kindred revisited.
	Haws DV, McKusick VA. Bull Johns Hopkins Hosp. 1963 Jul;113:20–30
Gene mapping	The anatomy of the human genome: a neo-Vesalian basis for medicine in the 21st century
	McKusick VA. JAMA. 2001 Nov 14;286(18):2289–2295
	The morbid anatomy of the human genome: the role of gene mapping in clinical medicine
	McKusick VA. Acta Paediatr Jpn. 1990 Jun;32(3):234–241
	The human genome through the eyes of Mercator and Vesalius
	McKusick VA. Trans Am Clin Climatol Assoc. 1981;92:66–90
Dysmorphology; the Birth Defects Conferences	The birth defects conferences: the last ten years and the next
	McKusick VA. Birth Defects Orig Artic Ser. 1978;14(6B):161–170
The 'Bar Harbor course'	40 years of the annual 'Bar Harbor Course' (1960–1999): a pictorial history
	McKusick VA, Naggert J, Nishina P, Valle D. Clin Genet. 1999 Jun;55(6):398–415
HUGO (Human Genome Organisation)	HUGO news. The Human Genome Organisation: history, purposes, and membership
	McKusick VA. Genomics. 1989 Aug;5(2):385–387
OMIM/MIM (Online/Mendelian Inheritance in Man)	Mendelian Inheritance in Man and its online version, OMIM
	McKusick VA. Am J Hum Genet. 2007 Apr; 80(4):588–604

major contribution was to chronicle the human gene map as it evolved and in particular develop the concept of 'the morbid anatomy of the human genome'.

Following the published record of progress in this field as documented by McKusick is not entirely straightforward, as the successive accounts overlap considerably. Some reports are in the published proceedings of the Human Gene Mapping Workshops held between 1973 and 1990 (themselves published in duplicate as part of both the *Birth Defects Original Articles Series* and as supplements to the journal *Cytogenetics and Cell Genetics*). Updated versions of the human gene map also appear as part of the introductory material for successive editions of McKusick's *Mendelian Inheritance in Man* and its online successor OMIM (see below). Perhaps the most definitive account is given by a series of four linked articles in JAMA (collected as [8]).

The linked concepts of the human gene map as a form of cartography and as the ultimate level of human anatomy are a further example of McKusick's historical approach, relating the topic on the one hand to Mercator and on the other to Vesalius (Fig. 16.1), as

Fig. 16.1 Mercator and Vesalius, representing cartography and anatomy, McKusick's two historical analogies for human gene mapping. (**a**) Gerard Mercator (1512–1594) and (**b**) Andreas Vesalius (1514–1564)

can be seen in his specifically historical articles on human gene mapping listed in Table 16.3. He admits that *'sometimes I have mixed the anatomic and cartographic metaphors'* [1].

The Human Genome Project, HUGO and 'Genomics'

By the late 1980s, the human gene map was already dense, and DNA technology had made it possible to move from mapping to actually isolating a human gene. Discussions about the systematic sequencing of the human genome had already begun, involving for the most part basic scientists completely outside medical genetics and largely unaware of the progress of the human gene map. Victor McKusick's role as a bridge between the two very different communities and cultures proved to be a most important one; at the contentious 1986 Cold Spring Harbor Symposium on the Human Genome, he gave a paper [9] on the 'status and prospectus' of the human gene map and, as he states [1], 'I think it was an eye-opener to molecular biologists'. It also provided an opportunity for resolving the dispute over how best to progress the Human Genome Project by taking a 'map first, sequence later' approach that was vindicated both by the rapid completion of the human gene map, mainly by the French Genethon group [10], and by the subsequent spectacular acceleration provided by new sequencing techniques, leading to the completed human genome sequence in 2003.

Alongside the project itself, McKusick was also much involved in developing the international organisational framework that would underpin it, the Human Genome Organisation (HUGO), of which he became first president, and also the new journal, Genomics. He describes this chapter of events in his retrospective article, *A 60-Year Tale of Spots, Maps, and Genes* [1], and also mentions it elsewhere [11].

Mendelian Inheritance in Man and OMIM

Along with the topic of human gene mapping, Victor McKusick's most enduring contribution has been the delineation of human Mendelian genetic disorders, beginning with his detailed studies of inherited disorders of connective tissue and the analysis of recessively inherited conditions in the Amish of Pennsylvania, progressing to the 'nosology' of birth defects as recorded in the Birth Defects Conferences volumes, and eventually becoming a definitive and all-encompassing documentation of single gene determined disorders, as the human gene map and his catalogues of inherited diseases evolved.

Typically, McKusick conceived and evolved a broad framework from the outset that could unify what might otherwise have become a fragmented and scattered series of contributions and which could integrate not just his own but other people's observations. Initially, this began as a 'journal club' to which his research fellows, mostly clinicians at that time, could bring material at evening meetings held in his home, these being assembled as annual reviews [12] over a 6-year period. The next step was to make as complete as possible a documentation for X-linked disorders, resulting in a remarkable but little known monograph, *On the X Chromosome of Man* [13], which was itself the predecessor to his definitive *Mendelian Inheritance in Man: A Catalog of Human Genes and Genetic Disorders*, whose first edition appeared in 1966 and the final (12th) print edition in 1998 [14].

Fig. 16.2 Successive print editions (1–12) of McKusick's *Mendelian Inheritance in Man* [31]. Each edition was of different colour

McKusick, in one of his last publications [15], has fortunately given us a detailed history of this monumental work and its online successor OMIM. He also emphasises its historical value, especially of the printed version, noting that:

> The twelve editions are serial cross sections of the field over the 33 years, 1966 to 1998, spaced three years apart, on average.

And understandably lamenting that:

> The historian in me regrets the loss of the archival function of the print edition.

Indeed, the printed volumes are historical documents in their own right (Fig. 16.2), their content developing mainly by accretion from one edition to the next, with very little rearrangement or deletion of material.

The decision, though, to make the catalogues computer-based from the outset, highly prescient at this early stage of using computers in biology and medicine, undoubtedly enhanced their value as a source for detailed analysis, given mainly in the front material of the printed volumes, and recording the numbers of genes and genetic disorders delineated and mapped, together with details of the 'morbid anatomy of the human genome', as already mentioned.

The continuously updated OMIM (Online Mendelian Inheritance in Man; http://www.ncbi.nlm.nih.gov/omim/) may have lost some of the historical value of the printed original, but it has undoubtedly ensured its lasting value and continuing development after McKusick's death and will remain the foundation for much of the practice and knowledge of medical genetics for the foreseeable future.

The 'Bar Harbor' Course

As a final example of important aspects of human and medical genetics that Victor McKusick has both created and chronicled, the 'Short Course in Medical Genetics' deserves a mention. Organised by Johns Hopkins faculty staff, with McKusick presiding, the 2-week course was held at The Jackson Laboratory, Bar Harbor, Maine, beginning in 1960; initially, it alternated with a mouse genetics course, but the two merged in 1979, and the course continues to flourish to this day. The combination of

Fig. 16.3 The Bar Harbor 'Short Course in Medical Genetics'. (**a**) Frank Ruddle lecturing at the Bar Harbor course. (**b**) Victor and Anne McKusick on the summit of Cadillac mountain, near Bar Harbor, Maine; sent as greeting card 2000

a high standard of faculty and presentations, with informality and a pleasant summer venue (Fig. 16.3), has given it a special and memorable character and has also generated a European counterpart, held for many years at Sestri Levante, Italy, with whose development McKusick was also closely involved. With around 100 students per year over a time span of more than 40 years, the Bar Harbor course has had a major influence on how medical genetics has developed.

The course has been documented in a series of articles [16, 17] and is also described in several of McKusick's wider reviews, but the 'pictorial history' of the course [18], illustrated by his own photographs, is of particular value and underlines the archival value of these images, as described below, despite (or more likely because of) them being informal 'snaps' of variable quality.

Biographical Articles

Most of Victor McKusick's historical publications relate to topics rather than to individuals, but there are several exceptions to this (Table 16.4). Two of his earliest papers [28, 30] relate to the nineteenth-century British physicians Jonathan Hutchinson and Frederick Parkes Weber, who had a particular interest in 'chronic diseases' (some of them genetic such as Hutchinson-Gilford progeria), a field which McKusick was to use as his base point in developing medical genetics when he took over the 'Moore Clinic' at Johns Hopkins in 1957. He was to return to Hutchinson's contributions at the end of his career in a paper published over 50 years later [19]. The original paper had been largely based on the superb collection of medical drawings compiled by Hutchinson, which after his death had been obtained for Johns Hopkins by William Osler and which had lain undisturbed in their packing cases in the library basement until McKusick and his then fiancée Anne unpacked and catalogued them.

Two other portraits are those of Marcella Boveri O'Grady (1863–1950) [20], her contributions having previously been overshadowed by those of her husband Theodor Boveri (1862–1915) whose landmark book on the genetic basis of cancer [21] she translated from German (a new translation has recently been produced by Henry Harris) [22]. The second article [23] is on Walter Sutton, who in 1902, while working as a student with Edmund Wilson, founder of cell biology, formulated the chromosome theory of heredity [24], a theme linking the two articles since Theodor Boveri had independently reached the same conclusions.

Table 16.4 Biographical articles

Marcella Boveri O'Grady	Marcella O'Grady Boveri (1865–1950) and the chromosome theory of cancer
	McKusick VA. J Med Genet. 1985 Dec;22(6):431–440
William Osler	Osler as a medical geneticist
	McKusick VA. Johns Hopkins Med J. 1976 Oct;139(4):163–174
Walter Sutton	Walter Sutton and the physical basis for Mendelism
	McKusick VA. Bull Hist Med. 1960; 35:487–497
Frederick Parkes Weber	Frederick Parkes Weber— 1863–1962
	McKusick VA. JAMA. 1963 Jan 5;183:45–49
Jonathan Hutchinson	The clinical observations of Jonathan Hutchinson
	McKusick VA. Am J Syph Gonorrhea Vener Dis. 1952 Mar;36(2):101–126
	The Gordon Wilson Lecture: The clinical legacy of Jonathan Hutchinson (1828–1913): syndromology and dysmorphology meet genomics
	McKusick VA. Trans Am Clin Climatol Assoc. 2005;116:15–38

Fig. 16.4 (**a**) William Osler working at his desk at Johns Hopkins Hospital (Courtesy of Victor McKusick). (**b**) The desk was also used by McKusick when he also was head of Medicine there (Courtesy of Alan Mason Chesney Archives)

Boveri died relatively young, soon after publishing his cancer monograph (and before Marcella had translated it). Sutton left the field of genetics, never submitting his Ph.D. thesis and becoming a surgeon, sadly dying of acute appendicitis in his 30s. These articles remind us of how many promising careers of scientists were cut off in their prime or before by sickness, or by war, with few attaining such a full span of active years as did McKusick himself.

A final person to be mentioned is William Osler, head of Medicine at Johns Hopkins and later regius professor of Medicine at Oxford University. McKusick was a fervent admirer of Osler and was proud also to be appointed head of Medicine, using Osler's original desk (Fig. 16.4) and office. His article, *Osler as a medical geneticist* [25], reminds us that Osler described a number of genetic disorders, even though he was not particularly interested in their hereditary aspects.

Other Historical Contributions

As shown by the quotation at the head of this article, and by what has already been described, Victor McKusick had a historical interest in all of the areas that he studied, visible from the beginning of his career and a constant feature throughout it. But in addition to the specific historical writings that I have outlined here, there are other parts of his historical legacy that are of equal importance.

Records Archive

One great advantage of McKusick's long and settled career at Johns Hopkins is that there were no moves to prompt disturbance or destruction of archival material. Not only was this true for his hospital-based records, but, even more important, his home, the location for much of his writing, remained unchanged until shortly before his death. Those privileged to have visited or stayed there will remember the strata of papers extending throughout much of the large house, with only Victor himself able to provide the key to their nature and location. Equally fortunate was that he was able to supervise their collection and removal to the Hopkins archive before he and Anne did finally move, leaving as complete an archive for future historical analysis as could reasonably be hoped for.

The archive (noted as being in 337 boxes!) is now deposited in the Alan Mason Chesney Medical Archives of the Johns Hopkins Medical Institutions (http://www.medicalarchives.jhmi.edu/papers/mckusick.html). It is currently listed as 'unprocessed', but fortunately, an online 'exhibit' (http://profiles.nlm.nih.gov/) has already been constructed as part of the US National Library of Medicine series 'Profiles in Science', which contains a vivid selection of correspondence, along with photographs (Fig. 16.5), research data, manuscripts, family papers and other materials. Already some historical research has used his records on the Amish hereditary disorder project [26], and there is no doubt that the archive will provide the foundation for decades of detailed historical research.

McKusick's photograph collection, already mentioned in relation to the 'Bar Harbor course', deserves a special mention. Thanks to his inveterate habit of photographing people and objects on every possible occasion, almost all informal and unposed, using a small and simple camera, we have a unique visual archive of medical geneticists and workers in allied fields over half a century (Fig. 16.5), ranging from the first Human Genetics Congress in Copenhagen in 1956 to very recent years, and often showing today's eminent and respectable leaders of the field at a time when they were callow youths! It is to be hoped that the many unidentified people in these photographs can be named before memories fade. Perhaps the best way of achieving this would be for the photos to be placed on the Web with a general invitation to identify individuals.

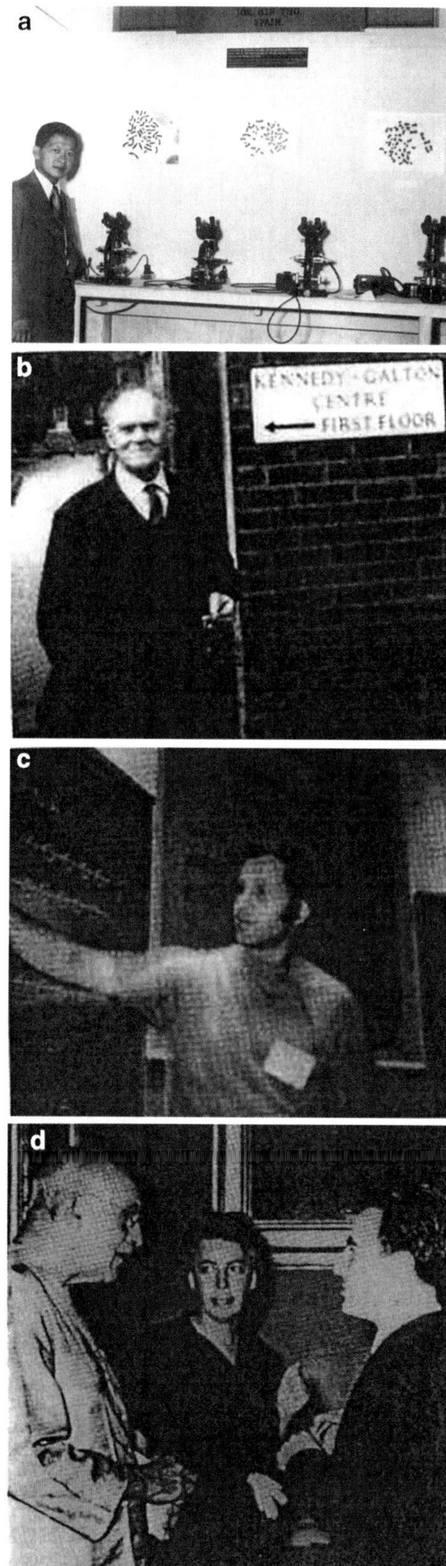

Fig. 16.5 Photographs by Victor McKusick as part of the history of medical genetics (Reproduced from [15], with permission). (**a**) J-H Tjio demonstrating the correct human chromosome number (46) at the first International Human Genetics Congress, Copenhagen, 1956. (**b**) Lionel Penrose, 1973, Photograph taken outside Kennedy-Galton Centre, London, shortly before Penrose's death. (**c**) Haig Kazazian at the Bar Harbor course and (**d**) JBS Haldane with his wife Helen Spurway and Marcello Siniscalco

Victor McKusick and History: Personal Aspects

As a person who played a major part in shaping and developing medical genetics, probably more than any other single individual, all aspects of Victor McKusick's life are of relevance to the history of medical genetics. His own writings, though, say relatively little about his life, apart from unfailingly, and rightly, acknowledging the total support of his wife Anne in all aspects of his work and wider life.

His 2006 article, *A 60-Year Tale of Spots, Maps, and Genes* [1], the closest to an autobiographical paper that he wrote, focuses almost exclusively on his work, and of his early upbringing, we learn only that:

> My identical twin brother Vincent and I grew up on a dairy farm in Maine in a family that valued scholastic achievement. Our elementary schooling (it was called grammar school) was in a one-room schoolhouse where the same teacher taught us seven of eight years.

He also tells us that he was influenced to enter medicine by developing, at the age of 15, an infectious abscess of the arm and axilla requiring prolonged hospital treatment, but cured by the new sulphonamide drug 'Prontosil'.

Fortunately, several recorded interviews have been made, notably one forming part of the American *Oral History of Human Genetics* series, of which the full transcript is available (http://ohhgp.pendari.com/). Here considerable detail emerges of his early life, though it is clear from the transcript that the interviewer required skill and persistence to bring him back to this from more professional aspects. A number of early photos are also given in the archive exhibit (Fig. 16.6).

Fig. 16.6 Photograph from the archived records of Victor McKusick (Courtesy of Alan Mason Chesney Archive, Johns Hopkins). The early 'Medical Genetics Journal Club' at the McKusick's home, 1958, with Anne McKusick and baby

Following McKusick's death aged 86 in 2008, a series of obituaries have appeared, though again mainly focused on professional aspects [45, 46, 47]. A potential source of later personal information (though unlikely to be historically accurate) would be the numerous anecdotes frequently recounted by fellows and others who have worked with him. It is greatly to be hoped that in due course, a detailed and objective scientific biography will be written.

Conclusion

Victor McKusick's numerous and varied contributions to the history of medical genetics have been a major step towards ensuring that the history of this key area of science and medicine is comprehensively documented. Historians of science and medicine may consider that his writings on the topic are biased by his intimate involvement with the field over the entire period of its existence; inevitably this is the case, and the historical record will undoubtedly benefit from wider contributions from others in medical genetics and from detailed analysis by professional historians, something that is just beginning to happen.

In the same way, McKusick's own role in the development of medical genetics will need the detached view of a skilled scientific biographer, not just the tributes of his many admirers. But, just as there can be no doubting that this role has been a central and pivotal one, so one can be equally certain that his historical contributions have provided a firm foundation on which others can build to ensure that the story of genetics in medicine worldwide is fully and accurately recorded, analysed and appreciated.

References

1. McKusick VA (2006) A 60-year tale of spots, maps, and genes. Annu Rev Genomics Hum Genet 7:1–27
2. McKusick VA (1992) Human genetics: the last 35 years, the present, and the future. Am J Hum Genet 50(4):663–670
3. Penrose LS (1967) Presidential address: the influence of the English tradition in human genetics. In: Crow JF, Neel JV (eds) Proceedings of the third international congress of human genetics. The Johns Hopkins University Press, Baltimore, pp 13–25
4. Harper PS (2008) A Short History of Medical Genetics. Oxford University Press, New York
5. McKusick VA (1989) Historical perspectives: the understanding and management of genetic disorders. Md Med J 38(11):901–908
6. McKusick VA (1978) The birth defects conferences: the last ten years and the next. Birth Defects Orig Artic Ser 14(6B):161–170
7. Donahue RP, Bias WB, Renwick JH, McKusick VA (1968) Probable assignment of the Duffy blood group locus to chromosome 1 in man. Proc Natl Acad Sci USA 61:949–955
8. McKusick VA (1988) The morbid anatomy of the human genome: a review of gene mapping in clinical medicine. Howard Hughes Medical Institute, Bethesda
9. McKusick VA (1986) The morbid anatomy of the human genome: a review of gene mapping in clinical medicine. Medicine 65(1):1–33
10. Weissenbach J, Gyapay G, Dib G et al (1992) A second-generation linkage map of the human genome. Nature 359:794–801

11. McKusick VA (1989) Forty years of medical genetics. JAMA 261(21):3155–3158
12. McKusick VA et al (1961) Medical genetics 1958–1960. An annotated review. Mosby, St. Louis
13. McKusick VA (1964) On the X Chromosome of Man. American Institute of Biological Sciences, Washington, DC
14. McKusick VA (1964–1998) Mendelian Inheritance in Man: A Catalog of Human Genes and Genetic Disorder. Johns Hopkins University Press, Baltimore, 1st edn 1966, 12th edn 1998
15. McKusick VA (2007) Mendelian inheritance in man and its online version, OMIM. Am J Hum Genet 80(4):588–604
16. Fuller JL, McKusick VA (1961) A short course in medical genetics. Eugen Q 8:98–101
17. Guethlein LA (1990) "The Bar Harbor Course": a 30-year veteran in the teaching of human genetics. Am J Hum Genet 46(1):192–207
18. McKusick VA et al (1999) 40 years of the annual 'Bar Harbor Course' (1960–1999): a pictorial history. Clin Genet 55(6):398–415
19. McKusick VA (2005) The Gordon Wilson lecture: the clinical legacy of Jonathan Hutchinson (1828–1913): syndromology and dysmorphology meet genomics. Trans Am Clin Climatol Assoc 116:15–38
20. McKusick VA (1985) Marcella O'Grady Boveri (1865–1950) and the chromosome theory of cancer. J Med Genet 22(6):431–440
21. Boveri T (1914) Zur Frage der Entstehung Maligner Tumoren (On the problem of the origin of malignant tumors) (trans: Boveri M). Jena, Fischer
22. Boveri T (2008) Concerning the Origin of Malignant Tumours (trans: Harris H). Cold Spring Harbor Laboratory Press, Woodbury
23. McKusick VA (1960) Walter Sutton and the physical basis for Mendelism. Bull Hist Med 35:487–497
24. Sutton WS (1903) The chromosomes in heredity. Biol Bull 4:231–251
25. McKusick VA (1976) Osler as a medical geneticist. Johns Hopkins Med J 139(4):163–174
26. Lindee S (2005) Moments of Truth in Genetic Medicine. The Johns Hopkins University Press, Baltimore
27. Haws DV, McKusick VA (1963) Farabee's brachydactylous kindred revisited. Bull Johns Hopkins Hosp 113:20–30
28. McKusick VA (1952) The clinical observations of Jonathan Hutchinson. Am J Syph Gon Vener Dis 36:101–126
29. McKusick VA (1958) Cardiovascular Sound in Health and Disease. Williams and Wilkins, Baltimore
30. McKusick VA (1963) Frederick Parkes Weber: 1863–1962. JAMA 183:45–49
31. McKusick VA (1966–1998) Mendelian inheritance in man: a catalog of human genes and genetic disorders. Johns Hopkins University Press, Baltimore, 1st edn 1966, 12th edn 1998
32. McKusick VA (1965) The royal hemophilia. Sci Am 213:88–95
33. McKusick VA (1973) Genetics and dermatology or if I were to rewrite Cockayne's *Inherited Abnormalities of the Skin*. J Invest Dermatol 60:343–359
34. McKusick VA (1975) The growth and development of human genetics as a clinical discipline. Am J Hum Genet 27(3):261–273
35. McKusick VA (1981) The last twenty years: an overview of advances in medical genetics. Prog Clin Biol Res 45:127–144
36. McKusick VA (1981) The human genome through the eyes of Mercator and Vesalius. Trans Am Clin Climatol Assoc 92:66–90
37. McKusick VA (1989) HUGO news. The Human Genome Organisation: history, purposes, and membership. Genomics 5(2):385–387
38. McKusick VA (1990) The morbid anatomy of the human genome: the role of gene mapping in clinical medicine. Acta Paediatr Jpn 32(3):234–241
39. McKusick VA (1991) Advances in medical genetics in the past 30 years. Res Publ Assoc Res Nerv Ment Dis 69:1–18

40. McKusick VA (1993) Medical genetics. A 40-year perspective on the evolution of a medical specialty from a basic science. JAMA 270(19):2351–2356
41. McKusick VA (1997) 1997 Albert Lasker Award for special achievement in medical science. Observations over 50 years concerning intestinal polyposis, Marfan syndrome and achondroplasia. Nat Med 3(10):1065–1068
42. McKusick VA (2001) The anatomy of the human genome: a neo-Vesalian basis for medicine in the 21st century. JAMA 286(18):2289–2295
43. McKusick VA (2004) Karl Landsteiner Award. From Karl Landsteiner to Peter Agre: 100 years in the history of blood group genetics. Transfusion 44(9):1370–1376
44. Rimoin DL, Connor JM, Pyeritz RE, Korf BR (2007) Emery and Rimoin's principles and practice of medical genetics. Philadelphia, Churchill-Livingstone
45. Hall JG, Victor A, McKusick MD (2008) A legend in his own time. Clin Genet 74:293–295
46. Rimoin DL, Victor A (2008) McKusick 1921–2008. Nat Genet 40:1037
47. Valle D (2008) Victor Almon McKusick, MD, 1921–2008, In: Memoriam. Am J Hum Genet 83:301–303

Chapter 17
The Influence of Victor A. McKusick on the Development of Training in Medical Genetics in Europe and in the World

Giovanni Romeo

Introduction

Among many other things, Victor was a born teacher and a living legend at Johns Hopkins Hospital where he spent all his professional life training students interested in medical genetics coming from all over the world. His scientific contributions on the development of modern medical genetics (and on its transition towards genetic medicine) have been so numerous and widespread that I have made the deliberate choice of focusing this chapter, using mainly my personal memories, on the influence he had within Europe (and elsewhere) through his teaching at the courses of the European School of Genetic Medicine (www.eurogene.org).

The Founding of the Twin Courses in Medical Genetics: Bar Harbor and Sestri Levante

I met Victor for the first time in 1968 at the annual course in Bar Harbor, Maine, which was already very popular because it had been training hundreds of practitioners and scholars. This course was first started in 1960, and I was there in 1968, while being a postdoctoral fellow at Hopkins. I attended the course together with a few other European participants, like Albert de la Chapelle and the late Giorgio Filippi. The course attracted a large number of health science faculty members and students from North America. One of the major objectives of the course was to upgrade the teaching of genetics, by "teaching the teachers." Indeed, a large proportion of people

---

G. Romeo, M.D.(✉)
University of Bologna Medical School and European Genetics Federation, via S. Petronio Vecchio, Bologna 42–40125, Italy
e-mail: romeo@eurogene.org

K.R. Dronamraju and C.A. Francomano (eds.), *Victor McKusick and the History of Medical Genetics*, DOI 10.1007/978-1-4614-1677-7_17,

Fig. 17.1 Sestri Levante—The first course in medical genetics (1988). The faculty included many Europeans and Americans, among whom you can see on the *first row* of the group picture, from *right* to *left*: Tom Caskey (in a *white jacket*), Jean-Claude Kaplan, Malcom Ferguson-Smith, Mansour Sarfarazi, Victor, Anne-Marie Fischauf, Jurg Ott, myself, Claudine Junien, Martin Bobrow, and Generoso Andria. In the *back rows*: H.H. Ropers, G. Uterman, and Kore Berg and in the *last row* Marcus Pembrey, Generoso Andria, and Andrea Ballabio

in genetics in the USA have attended the Bar Harbor course at some stage of their careers. The multiplier effect of this "teaching of teachers" was a catalyst in the development of medical genetics as a specialty in the USA and in Europe.

The cultural imprinting left on me by this course, as well as by my training at Hopkins, led me 20 years later to the idea that we should start a parallel course in Europe. Victor was very supportive of this idea, and we started the first European course in medical genetics in 1988 in collaboration with many European medical geneticists who were to become the leaders of the newly reorganized European Society of Human Genetics a few years later, in the early 1990s (Fig. 17.1).

This first course in medical genetics was a great success, and for many of us (including Victor), it was a sort of "love at first sight" for the venue, for the students coming from all over Europe, and for the idea itself of a "Bar Harbor twin course." The course took place every year until 2001 in Sestri Levante, a charming small town of the Eastern Riviera Ligure, close to Genoa, where I was head of Medical Genetics at the Gaslini Institute, the largest pediatric hospital in Italy, and professor of human genetics at the university.

One of the very successful social features of the Sestri Levante course was a beautiful excursion through the Cinque Terre where Victor and Anne would often be seen leading a group of students (Fig. 17.2).

Fig. 17.2 Excursion through the Cinque Terre medical genetics course 1988

During the course, Victor would always be sitting in the first row with many other faculty and moderating all the discussions. The tradition of the school required that faculty stay in residence and attend all lectures so that, as Victor would say, "they will keep each other honest," meaning that the quality of lectures is very much dependent on the spirit of peer control among faculty members. I believe this has been the key of the success of this course, which every year was the "best ever" as Victor would say (he would say the same for the Bar Harbor course). Overall, the students have always been extremely good and highly motivated but sometimes too shy to ask questions during plenary sessions. To circumvent this psychological block, we invented the "McKusick's box" where students could drop their questions anonymously (if they wanted to do so) addressed to individual speakers who would answer them at the end of each morning plenary session. Victor was excellent in guiding all the discussions without asking too many questions himself (he would have plenty), but from time to time, he could not resist the temptation of dropping his own questions in the "McKusick's box," as I could tell from his handwriting.

The Hybrid Courses and the "Cloning" of the Course in Medical Genetics

In more recent years, the use of written questions became a feature of the "hybrid courses" of the school, where students from remote training centers can ask questions to speakers by e-mail after following the lectures through live webcasting.

The tradition of streaming the plenary lectures of the courses started in the year 2000, at the last course in medical genetics held in Sestri Levante, but did not become a routine feature until 2005 when a grant from the European Commission through the EUMEDIS (Euro-Mediterranean Information Society) program to the European Genetics Foundation (which manages the courses of the European School of Genetic Medicine—*ESGM*) stimulated the development not only of a Euro-Mediterranean network for medical genetics but also the development of a program of distance learning called "hybrid courses." This development made it possible to "clone" the original course in medical genetics into many other courses, some taking place every year, others less often (Tables 17.1 and 17.2). ESGM courses are now usually webcast live to remote training centers (RTC), located in hospitals, clinics, universities, research institutes all over Europe, the Southern Mediterranean region, and the Middle East (Fig. 17.3). At each RTC, ESGM courses are thus followed live and discussion sessions and workshops are organized by local RTC faculty. In order to maximize interactive learning during these sessions, student questions are collected and e-mailed to the main training center and the answers are discussed and disseminated again by webcasting at the end of each morning plenary session. Hybrid courses have fostered the birth of several local teaching experiences in the respective national languages (in Spain, Morocco, Slovenia, Turkey, Iran, Cyprus, Malta, etc.). In these RTC, the morning lectures from the main training center represent at least half of the contents of the "hybrid courses," the other half being represented by workshops and discussion groups on the themes of the morning lectures led by local faculty.

At the medical genetics courses, Victor would always get excited by the number and quality of questions arriving from remote training centers all over the world. The idea of "cloning" the medical genetics course into many other courses, leaving the latter unaltered according to the initial format, arose during a discussion with

Table 17.1 ESGM courses repeated two or more times during the 2000–2011 period

2000–2011 ESGM courses	Editions	Students and faculties
Total (including Tables 17.1 and 17.2)		6,205
Bioinformatics	11	494
Medical Genetics	12	1,679
Integration of Molecular Cytogenetics, Microarrays, and Massive Sequencing	11	876
Genetic Counseling in Practice	10	361
Statistical Genetic Analysis of Complex Phenotypes	8	437
Cancer Genetics	7	581
Developmental Genes and Clinical Dysmorphology	6	237
Thalassemia and Related Disorders in the Mediterranean	5	205
Neurogenetics	3	152
Eye Genetics	2	93
Peer Review in Theory and Practice	2	71
Microbial Bioremediation and Environmental Genomics	2	74
Mitochondrial Medicine	2	162

Table 17.2 Courses which took place only once during the 2000–2011 period

2000–2011 ESGM courses held as single editions	16	783
Genetics, Medical Technologies, and Public Health Policies	1	24
Cardiogenesis and Congenital Cardiopathies: From Developmental Models to Clinical Applications	1	41
Comparative Health: The Reforms of the Health Care Systems in a Globalized World	1	29
Dal Test Genetico alla Chirurgia Profilattica: Workshop sui Problemi Etico-Giuridici Connessi alla Gestione del Carcinoma Mammario Ereditario	1	43
Developing Advanced Counseling Skills in Genetics	1	45
EMBO Practical Course on Networks in Biology Analysis, Modeling, and Reverse Engineering	1	75
Genetic Epidemiology in Chronic Kidney Disease Research	1	38
Genetics of Rheumatic and Auto-inflammatory Diseases	1	51
Genetics, Ethics, and the Law	1	41
Genomic and Cultural Evolution of Humans	1	42
Minicorsi sulle Malformazioni Congenite: Corso su Genetica e Clinica delle Cardiopatie Congenite	1	54
Next-Generation Sequencing for Rare and Common Genetic Disorders (2011)	1	96
Practical Systems Biology: Visualization and Reverse Engineering Gene Regulatory Networks	1	21
Prenatal and Postnatal Clinical Cytogenetics	1	44
SAFE Practical Workshop on "Noninvasive Prenatal Diagnosis"	1	95
Technologies and Genomics of Microbial Hydrogen Production	1	44

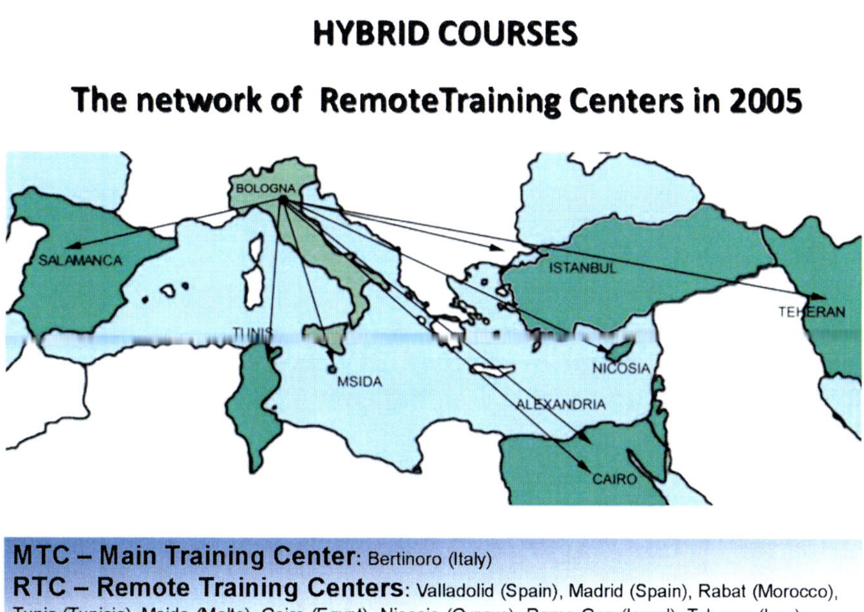

Fig. 17.3 The first network of hybrid courses in 2005

Victor on the future of the school. He was convinced that training in genetic medicine should be developed by combining the expertise of geneticists with that of clinical specialists from different fields of medicine. He was therefore suggesting that the European School should start organizing courses in neurogenetics, genodermatoses, ophthalmogenetics, etc. The most successful translation into practice of this idea of Victor has occurred during the last 2 years (2010–11) with the course entitled Eye Genetics which has mixed very well the competences of molecular and clinical geneticists with that of ophthalmologists (see www.eurogene.org for the scientific program of this course).

In 2002, the course in medical genetics, which had been going on in Sestri Levante for 14 years and the entire school had to move for logistic reasons to a new venue, Bertinoro di Romagna, located 80 km southeast of Bologna, where in the meantime I had moved with a group of my collaborators from Genova and Lyon. Victor was the director of the course in medical genetics in Bertinoro for the last time in May 2007. A few days later, upon his return to Baltimore, he was diagnosed to have cancer. He knew very well what he was facing and sent to all his friends an e-mail which started with the following words: "My renegade chromosomes…."

On May 12, 2007, at the end of the 20th course in medical genetics, the University of Bologna awarded an honorary degree in medical biotechnologies to Victor and to Mario Capecchi (it was, I believe, the 24th honorary degree for Victor). The "doctoral lectures" delivered by Victor and Mario on that occasion are part of the booklet entitled "Homage to Victor A. McKusick" (accessible through www.eurogene.org) which collected photographs, memories, and comments of a number of alumni and faculties who participated in the courses of the first 20 years of the European School of Genetic Medicine. These testimonials represent in my opinion one of the most significant documents of Victor's invaluable contribution to the education of so many generations of medical geneticists in Italy, in Europe, and in the world.

While Mario Capecchi won the Nobel Prize for medicine a few months later, Victor was conferred by the emperor of Japan the prestigious Japan Prize on April 2008, 3 months before his death. It was a memorable occasion for Victor, which he joyfully partook together with his family and a few Hopkins colleagues as documented by the video you can find at www.japanprize.jp.

The Euro-Mediterranean Center for Genetic Medicine Dedicated to Victor in Ronzano, Bologna

In order to expand the number of courses of the school and as a consequence of the "cloning" of the medical genetics course suggested by Victor, we needed a new venue not far from the city of Bologna where the European Genetics Foundation had established itself after my return in 2001 to my Alma Mater Studiorum. Such a venue was found at the Hermitage of Ronzano, on the hills of Bologna, 3.5 km from its medieval center. The owners of the hermitage are the "Servants of Mary," a monastic order very closely related to the Augustinians. They knew very well that the founder of genetics, Gregor Mendel, was an Augustinian monk, and they not only did rent to us a wing of their hermitage (which is the present provisional venue of the school) but they wanted also to donate to the European Genetics Foundation a huge piece of land (16,500 square meters) next to the hermitage where eventually the permanent venue of the school will be built, when we will succeed in raising the necessary amount of money.

The "*Giuseppe Levi and Victor McKusick Euro-Mediterranean Center for Genetics and Medicine*" was inaugurated in Ronzano on May 13, 2005. Victor himself was present together with the rector of the University of Bologna, Pier Ugo Calzolari; the president of Johns Hopkins University, William Brody; the mayor of Bologna, Sergio Cofferati; and many other city authorities. The center started to function as the provisional venue of the European School of Genetic Medicine in June 2007, after all the necessary equipment, including a large informatics laboratory donated by the Melò family, was in place. The name of Giuseppe Levi (1872–1965) was coupled with that of Victor to remember the most brilliant Italian medical scientist who, between the two World Wars, built in Turin a research institute where, among others, three of his students (Salvador Luria, Renato Dulbecco, and Rita Levi Montalcini) started their training. While Levi had to leave Italy in 1938 because of the racial laws passed by the fascist regime, his three students migrated one after the other to the USA and eventually got their Nobel Prize in different fields of medicine. Victor developed a great admiration for Giuseppe Levi whose memory and accomplishments had been totally ignored in Italy for 50 years. Renato Dulbecco, who could not attend the inauguration of the center dedicated to his mentor as well as to Victor, wrote in a letter dated May 2005: "I think that Levi's attitude explains why the three of us ended up, many years later, earning the Nobel Prize in different fields. He had taught us the right attitude for doing research."

In the world, there are two more training centers or institutes of genetic medicine that bear the name of Victor McKusick, besides the *Levi-McKusick Euro-Mediterranean Center*. The first one is the McKusick-Nathans Institute of Genetic Medicine in Baltimore, dedicated to Victor and Daniel Nathans by their own alma mater. Daniel Nathans was professor of molecular biology at Hopkins and was awarded the Nobel Prize in 1978 for discovering restriction enzymes, which are at the basis of genetic engineering. The other one (the McKusick-Zhang Center) was founded at the Peking Union Medical College in the year 2000 by Wilson Lo, a medical geneticist who was a resident in medicine at Hopkins when Victor was chief resident.

The "Internationalization" of Genetic and Genomic Training as Victor's Main Legacy

The existence of three institutes or centers named after Victor in three different continents reflects one the most significant achievement he made over his long and productive career, namely, the training of several generations of geneticists throughout the world. As already said, Victor started in 1960 the first course in medical and mammalian genetics in collaboration with the Jackson Laboratory in Bar Harbor, Maine. Known officially as "The Short Course on Medical and Experimental Mammalian Genetics" and unofficially as the "Short Course," this collaborative effort between the Jackson Laboratory and Johns Hopkins continues to the present and has over the last 51 years trained >5,000 students and involved >250 faculty about one-third of which come from academic centers other than the two sponsoring institutions. The Short Course also served as the model for the courses offered by the European School of Genetic Medicine, now taking place in Ronzano, Bologna, as just described. Finally, Victor was also instrumental in establishing in 2002 an annual course in Human Genetics sponsored by the McKusick-Zhang Center in Beijing and directed by Ethylin Jabs from Mt. Sinai Medical Center and Kung-Yee Liang from Johns Hopkins School of Public Health.

In July 2008, until the very last day of his life, Victor was following from his home in Baltimore all the lectures of the 49th Short Course which he could not attend in Bar Harbor and which were streamed just for him.

I believe that the best way to take up Victor's legacy for training in genetics and genomics is to disseminate further the courses that he initiated and/or inspired by using the most popular communications tools that Internet makes available today. In order to set up a strategy to reach this goal, we, as human and medical geneticists, have to realize that the international expansion of the contents of these courses, using state-of-the-art communication technology, can go a long way towards meeting the demand for training in genetics and genomics around the world. The expansion over the last 12 years of the hybrid courses (Figs. 17.3 and 17.4) and the total number of courses of the European School of Genetic Medicine during the same period (Figs. 17.5–17.7 and Tables 17.1 and 17.2) are the proofs that this "internationalization" is not only possible but is also very much needed today, as Victor himself was suggesting. Looking at the breakdown of participants to courses of the European School of Genetic Medicine, it is clear in which fields most of the demand for training in genetics and genomics resides, starting with medical genetics and bioinformatics (Fig. 17.5). At the same time, training in new "niches" is becoming increasingly important as shown, for example, by the course in "Next-Generation Sequencing" which took place for the first time in Ronzano in 2011 and was very successful in terms of participants (see Table 17.2).

The international nature of the European School is also evident if one looks at the breakdown of students during the last 12 years: 33% are from outside Europe (Fig. 17.6) while, among Europeans, the Italians do not go over 35% (Fig. 17.7), an estimate which includes some courses in Italian, not attended by non-Italian students (see Table 17.2). From Tables 17.1 and 17.2, it is also clear that the school has been increasing the number of courses offered over the years. This increase happens by trial and error, that is, by selecting the courses which attract most students. This is because all the courses must be self-supporting since in recent years the financial support from international funding agencies for this type of training activities has been steadily decreasing. It is also worth underlining the strong participation of

ESGM Students (2005-2011) - Hybrid Courses

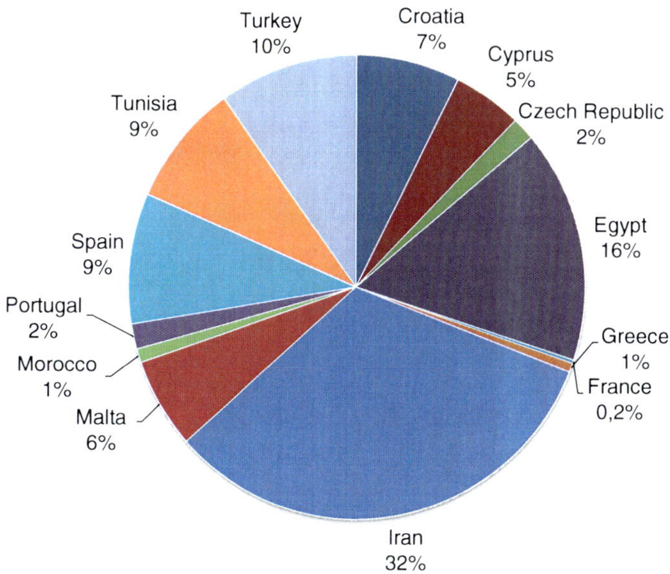

Croatia	92
Cyprus	61
Czech Republic	20
Egypt	203
France	3
Greece	8
Iran	407
Malta	79
Morocco	12
Portugal	22
Spain	114
Tunisia	107
Turkey	125
Total	**1253**

Fig. 17.4 Breakdown per country of students attending hybrid courses during the 2005–2011 period

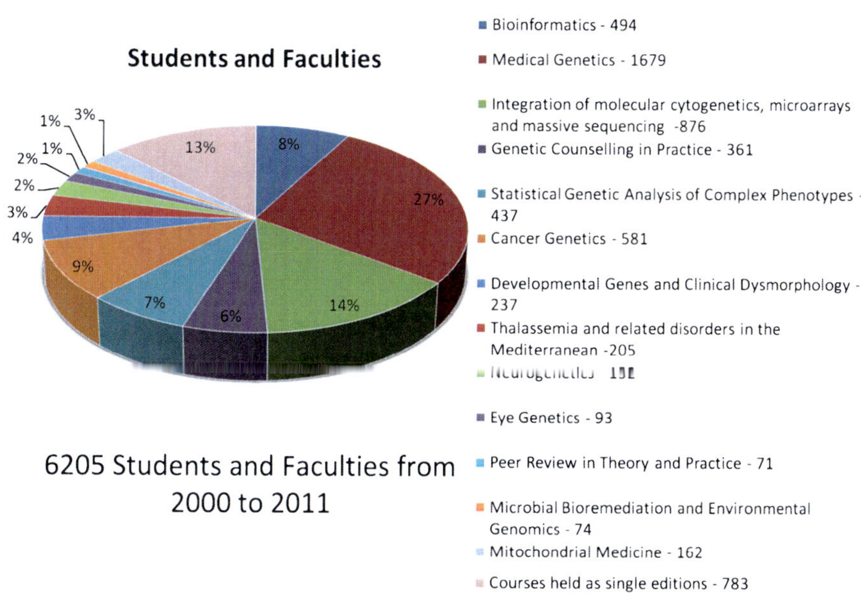

Fig. 17.5 Breakdown of the 6,205 students and faculties participating in courses of the European School of Genetic Medicine between 2000 and 2011. Some courses took place only once, while the most successful ones were repeated two or more times during the 2000–2011 period like medical genetics with 12 editions (see Tables 17.1 and 17.2)

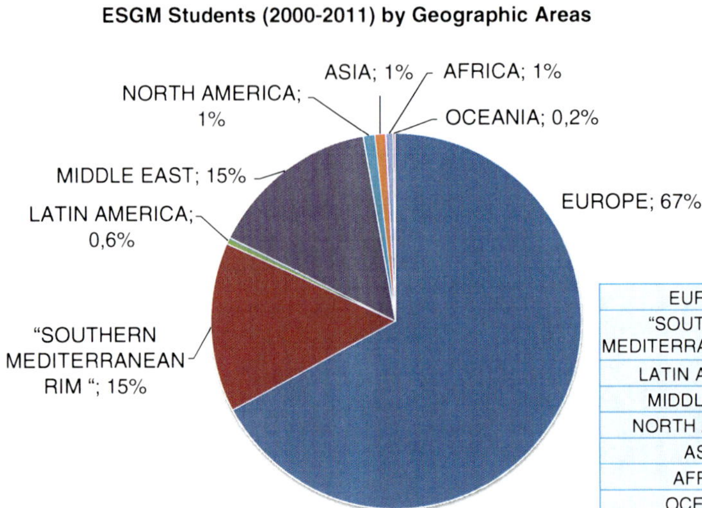

EUROPE	3983
"SOUTHERN MEDITERRANEAN RIM"	868
LATIN AMERICA	33
MIDDLE EAST	883
NORTH AMERICA	60
ASIA	56
AFRICA	36
OCEANIA	10
TOTAL	**5929**

Fig. 17.6 Breakdown by geographic area of only the 5,929 students (including RTC students)

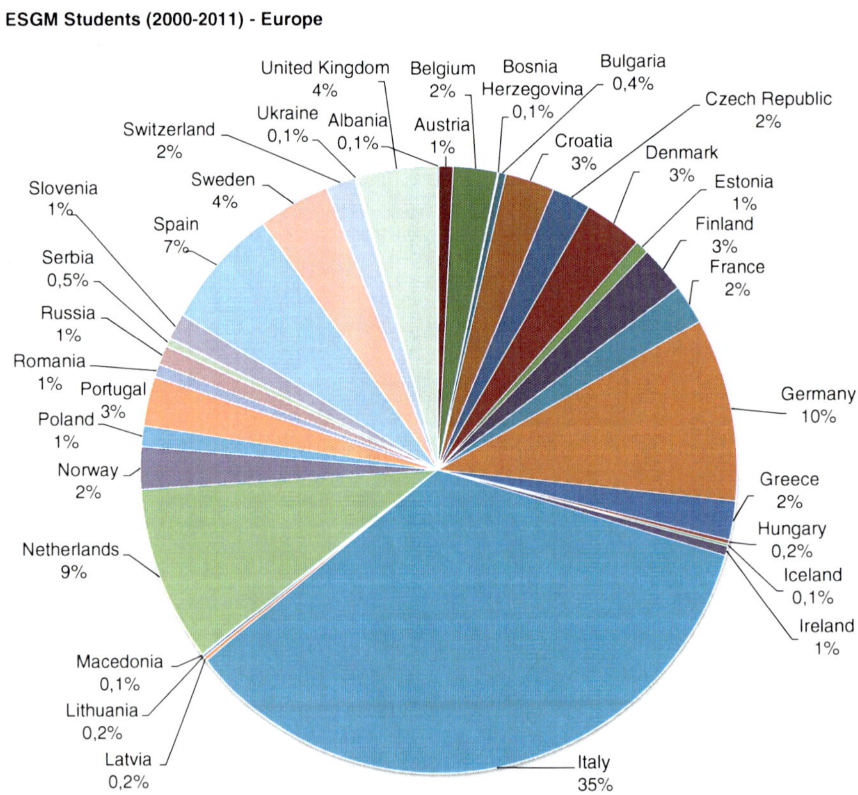

Fig. 17.7 Breakdown by nationality of only the 3,923 European students (including RTC students)

students from the Southern Mediterranean rim and the Middle East, the two geographic regions which have mostly benefited from the ESGM courses (Fig. 17.6).

In conclusion, I propose to take up Victor's legacy for training in genetics and genomics and suggest that professional and scientific international societies (not only those of human and medical genetics) build together a common program in training in genetics and genomics suited to their needs. A *board of directors* (BoD) where every interested society should have a representative will become the tool to manage this international initiative and to guarantee the necessary control of the overall administration of the courses. This initiative which I suggest to call "*International Training Program in Genetics and Genomics in Memory of Victor A. McKusick*" will take advantage of the experience, the competences, and the logistics developed during the past 24 years by the European School of Genetic Medicine. The characterizing features of such a program will be (a) the high quality of faculty teaching in each course, to be monitored by the BoD which will appoint the Directors of the courses, (b) the promotion of satellite training experiences in peripheral locations through live streaming of the courses following the model of hybrid courses initiated by the European School of Genetic Medicine, (c) the availability as open-access material of the videos, PowerPoint slides, and abstracts used in the different courses following the model developed by the EU funded project "Eurogene" (see: www.eurogene.eu).

Every person or scientific society interested in joining this effort can contact me at: romeo@eurogene.org. The initiative should be self-supporting through fund-raising and registration fees of participants to the courses. The first individual donation is represented by the check accompanying the Award for Excellence in Human Genetics Education which the American Society of Human Genetics wanted to bestow on me in Montreal on October 12, 2011, during the 12th International Congress of Human Genetics. It is a great honor for me not only to have been chosen by my colleagues as the recipient of this award for the year 2011 but also to have the possibility to dedicate it to the memory of Victor McKusick and to initiate the fund-raising campaign for this program.

Acknowledgments I am very grateful to Dr. Anne McKusick for correcting some of my personal memories; to Mr. Matteo Dutto, my collaborator at EGF, for preparing the data presented in Fig. 17.1 through 17.7 and Tables 17.1 and 17.2; and to all my colleagues who have been teaching at the European School of Genetic Medicine (in particular to Drs. Rando Allikmets: rla22@columbia,edu and Bart Leroy) for very useful discussions and very useful suggestions on how to set up the "International Training Program in Genetics and Genomics in Memory of Victor A. McKusick." A special acknowledgment is due to the March of Dimes and to the European Society of Human Genetics for their support to the courses of the European School of Genetic Medicine.

Note added in proof As I anticipated while writing this article, the survival of the European School of Genetic Medicine is jeopardized by the heavy economical crisis which is hitting Italy and Europe in general. Under these conditions EGF had to reduce already the number of coursers of the school and will have to close, at least for the time being, the temporary venue of the "G. Levi and V.A.McKusick EuroMediterranean Center for Genetics and Medicine" in Ronzano-Bologna. Under these conditions I want to renew my appeal to all the persons interested in the idea of internationalization of advanced training in genetics/genomics, to get in touch with me in order to find possible solutions and collaborations capable of launching as soon as possible the Board of Directors described in the text.

Appendix A
Bibliography of Victor A. McKusick

1. McKusick VA (1949) Broedel's ulnar palsy, with unpublished Broedel sketches. Bull Hist Med 23:469–479
2. Haines JS, Grabstald H, McKusick VA, Rodger RC, Jones RM (1949) Carcinoma of lung simulating primary carcinoma of urinary bladder. Urol Cutan Rev 53:324–327
3. Jeghers H, McKusick VA, Katz KH (1949) Generalized intestinal polyposis and melanin spots of the oral mucosa, lips and digits. New Eng J Med 241:993–1005, 1031–1036
4. Southworth JL, Peirce EC II, Rawson FL Jr, McKusick VA (1950) The state of the circulation in ventricular fibrillation, Surgical forum. Saunders, Philadelphia, pp 221–226
5. Jeghers H, McKusick VA, Katz KH (1950) The syndrome of generalized intestinal polyposis and melanin spots of the oral mucosa, lips and digits. Ciba Symp 2:199
6. Truitt EB Jr, McKusick VA, Krantz JC Jr (1950) Theophylline blood levels after oral, rectal and intravenous administration, and correlation with diuretic action. J Pharmac Exp Therap 100:309–15100:309–315
7. McKusick VA (1950) Effect of lithium on the electrocardiogram of animals. Fed Proc 9:84
8. Truitt EB Jr, McKusick VA (1950) Theophylline blood levels after oral, rectal and intravenous administration of aminophylline. Fed Proc 9:322
9. McKusick VA (1950) Syllabus: treatment of poisoning. Mimeographed 28 Apr 1948; revised
10. Southworth JL, McKusick VA, Peirce ECH, Rawson FL Jr (1950) Ventricular fibrillation precipitated by cardiac catheterization; complete recovery of the patient after forty-five minutes. JAMA 143:717–720
11. McKusick VA (1950) Electrokymography in constrictive pericarditis: comparisons with roentgenky mography and right ventricular pressure curves. In: Proceedings first Conference on Electrokymography, 25–26 May 1950 pp 125–137
12. Stimson WH, McKusick VA (1951) Febrile reactions to quinidine. Am J Med Sci 221:440–443
13. Callaway JJ, McKusick VA (1951) Carbon-dioxide intoxication in emphysema: emergency treatment by artificial pneumoperitoneum. New Eng J Med 245:9–13
14. Fisher AM, McKusick VA (1952) Bacteroides infections: some clinical and therapeutic features. Trans Am Clin Climatol Assoc 64:133–4564:133–145
15. McKusick VA (1952) Chronic constrictive pericarditis. I. Some clinical and laboratory observations. Bull JHH 90:3–26
16. McKusick VA (1952) Chronic constrictive pericarditis. II. Electrokymographic studies and correlations with roentgenkymography, phonocardiography, and right ventricular pressure curves. Bull JHH 90:27–41
17. McKusick VA, Cochran TH (1952) Constrictive endocarditis: report of a case. Bull JHH 90:90–97

K.R. Dronamraju and C.A. Francomano (eds.), *Victor McKusick and the History of Medical Genetics*, DOI 10.1007/978-1-4614-1677-7,

18. Shaffer CF, Chapman DW (1952) Correlative cardiology: an integration of cardiac function and the management of cardiac disease. Bull JHH 91:503–505, Review
19. McKusick VA, Stacey CH (1952) Depigmentation from adhesive tape. JAMA 150:1332
20. Stumpf P (1952) Kymographische Roentgendiagnostic zur Beureteilunt in Beispielen. Bull JHH 90:332–333, Review
21. Berman C (1952) Primary carcinoma of the liver: a study in incidence, clinical manifestation, pathology and aetiology. Bull JHH 91:504–505, Review
22. McKusick VA, Grob D (1952) Report of three new cases of parathion poisoning. Quarterly Progress Report. Chemical Warfare Service Contract
23. Scarborough WR, McKusick VA, Baker BM Jr (1952) The ballistocardiogram in constrictive pericarditis before and after percardiectomy. Bull JHH 90:42–63
24. McKusick VA (1952) The clinical observations of Jonathan Hutchinson. Am J Syph Gon 36:101–126
25. McKusick VA (1952) A year in Osler, being an account of some clinical Experience of the Osler House staff, Johns Hopkins Hospital, from July 1, 1951 to June 30, 1952, 90 pages. Duplicated, print shop, Johns Hopkins Hospital
26. Ross RS, McKusick VA (1953) Aortic arch syndromes: diminished or absent pulses in vessels arising from the arch of the aorta. Arch Int Med 92:701–740
27. Fisher AM, McKusick VA (1953) Bacteroides infections: clinical, bacteriological and therapeutic features of fourteen cases. Am J Med Sci 225:253–273
28. McKusick VA (1953) Boeck's sarcoid of the stomach, with comments on the etiology of regional enteritis. Gastroenterology 23:103–1323:103–113
29. Graybiel A, White PD (1953) Electrocardiography in practice. Bull JHH 92:181–82, 92:181–182, Review
30. Kemp T (1953) Genetic and disease. Bull JHH 92:450–51, 92:450–451, Review
31. Heckmann K (1953) Gruandise der Elektrokymographie: Phasenanalyse Raumkymographie des Herzens. Bull JHH 92:348, Review
32. Gray J (1953) History of the royal medical society, 1737–1937. Bull JHH 92:183, Review
33. McKusick VA (1953) Numerous abstracts of articles on cardiovascular topics. Circulation 7–8
34. Von Oettigen WF (1953) Poisoning: a guide to clinical diagnosis and treatment. Bull JHH 92:77, Review
35. McKusick VA (1953) Virchow-Troisier node: an occasional conspicuous manifestations of gallbladder cancer. South Med J 46:965–6746:965–967
36. McKusick VA (1953) The diagnosis of organic mitral stenosis in the presence of sickle-cell anemia. Am Heart J 46:467–7546:467–475
37. McKusick VA (1953) The minutes of the Johns Hopkins Medical History Club, 1890 to 1894: reflections on the role of the organization. Bull Hist Med 27:177–8127:177–181
38. Williams TF, McKusick VA (1954) Bernhardt Bang: physician, veterinarian scientist (1842–1932). Bull Hist Med 28:60–72
39. Rapport SB, Wright H (1954) Great adventures in medicine. Am Rev Tubercul 69:129, Review
40. McKusick VA (1954) Numerous abstracts of articles on cardiovascular topics. Circulation 9 and 10
41. McKusick VA, Hahn DP, Brayshaw JR, Humphries JO (1954) Some hemodynamic effects of the Hufnagle operation for aortic regurgitation: studies in models and a patient. Bull JHH 95:322–3795:322–337
42. McKusick VA, Talbot SA, Webb GN (1954) Spectral phonocardiography: studies of heart sounds and murmurs with a modified Bell sound spectrograph. J Clin Invest 33:954–55, 33:954–955, Abstract
43. McKusick VA, Talbot SA, Webb GN (1954) Spectral phonocardiography: application of the bell sound spectrograph to phonocardiography. Fed Proc 13:522
44. McKusick VA, Talbot SA, Webb GN (1954) Spectral phonocardiography: problems and prospects in the application of the bell sound spectrograph to phonocardiography. Bull JohnHopkins Hosp 94:187–9894:187–198

45. McKusick VA, Webb GN, Bradshaw JR, Talbot SA (1954) Spectral phonocardiography: clinical studies. Bull JHH 95:90–110
46. Grob D, Johns RJ, McKusick VA, Harvey AM (1954) Studies on the stimulating and depressant effects of acetylcholine on neuromuscular function in man. J Clin Invest 33:939, Abstract
47. Bancroft H, Swan HJC (1954) Sympathetic control of human blood vessels. Bull JHH 94:113–1494:113–114
48. McKusick VA (1954) The Hutchinson collection. J Assoc Med Illustrators 6:63
49. McKusick VA (1954) The cardiovascular and genetic aspects of Marfan's syndrome, a heritable disorder of connective tissue. Bull JHH 94:159–6194:159–161
50. Pinner B, Miller BF (1954) When doctors are patients. Bull Hist Med 28:497–828:497–498
51. McKusick VA (1954) The effect of lithium on the electrocardiogram of animals and the relation of this effect to the ratio of the intracellular and extracellular concentrations of potassium. J Clin Invest 33:598–1033:598–610
52. McKusick VA (1954) The study of mitral regurgitation by roentgen kymography, with observations on the movement of cardiac calcification. Am J Roentgenol 71:961–7871:961–978
53. McKusick VA, Kay JH, Isaacs JP (1955) Constrictive pericarditis following traumatic hemopericardium. Ann Surg 142:97–103
54. McKusick VA, Harvey AM (1955) Diseases of the pericardium. Adv Intern Med 7:157–200
55. McKusick VA, Cooley RD (1955) Drainage of right pulmonary vein into inferior vena cava: report of a case, with radiological analysis of the principal types of anomalous pulmonary venous return. New Eng J Med 252:291–01252:291–301
56. Grob D, McKusick VA (1955) Effect of anti-hypertensive drugs on renal blood flow and function. Bull JHH 96:164
57. McKusick VA (1955) Heritable disorders of connective tissue. I. The clinical behavior of hereditary syndromes. J Chronic Dis 2:491–992:491–499
58. McKusick VA (1955) Heritable disorders of connective tissue. II. The biology of normal connective tissue. J Chronic Dis 2:500–072:500–507
59. McKusick VA (1955) Heritable disorders of connective tissue. III. The Marfan syndrome. J Chronic Dis 2:609–442:609–644
60. Holldack K (1955) Lehrbuch der Auskultation und Perkussion. Bull JHH 97:62–63
61. McKusick VA, Murray GE, Peeler RG, Webb GN (1955) Musical murmurs. Bull JHH 97:136–7697:136–176
62. McKusick VA (1955) Numerous abstracts of articles on cardiovascular topics. Circulation 11 and 12
63. McKusick VA, Webb GN, Humphries JON, Reid JA (1955) On cardiovascular sound: further observations by means of spectral phonocardiography. Circulation 11:849–7011:849–870
64. McKusick VA, Webb GN, Vescia FG (1955) Phonocardiographie spectrale: une nouvelle methode d'analyse des bruits cardiovasculaires. Arch Mal Coeur 48:505–1248:505–512
65. McKusick VA (1955) Primordial dwarfism and ectopia lentis. Am J Hum Gen 7:189–987:189–198
66. McKusick VA (1955) Rheumatic restenosis of the mitral valve: report of a case with death over five years after valvulotomy. Arch Int Med 95:557–6295:557–562
67. McKusick VA (1955) Spectral phonocardiography. (Letter) Lancet 2:677
68. McKusick VA, Kline EW, Webb GN (1955) Spectral phonocardiographic demonstrations of selected varieties of cardiovascular sounds. Am Heart J 49:911–3349:911–933
69. McKusick VA, Jenkins JT, Webb GN (1955) The acoustic basis of the chest examination: studies by means of sound spectrography. Am Rev Tubercul 72:12–34
70. McKusick VA (1955) The cardiovascular aspects of Marfan's syndrome: a heritable disorder of connective tissue. Circulation 11:321–4211:321–342
71. McKusick VA (1955) The early history of Parkman. Piscataquis Observer (Dover-Foxcroft, Me.)
72. McKusick VA (1955) The illnesses of the great and near-great. Bull Hist Med 29:377–8129:377–381

73. McKusick VA (1955) The new look in medical genetics. Clin ResProc 3:187, Editorial
74. McKusick VA, Reagan WP, Santos GW, Webb GN (1955) The splitting of heart sounds: a spectral phonocardiographic evaluation of clinical significance. Am J Med 19:849–6119: 849–861
75. McKusick VA (1955) Clinical, genetic and laboratory investigation of three types of hereditary dystrophy of connective tissue. Year Book of the American Philosophical Society: 205–08
76. McKusick VA, Murray GE, Peeler RG, Webb GN (1955) Musical cardiovascular murmurs. Mod Med: 81–83
77. McKusick VA Carcinoid cardiovascular disease. Yearbook of cancer, 1956–1957 series, pp 93–97
78. Crowe FW, Schull WJ, Neel JV (1956) A clinical pathological and genetic study of multiple neurofibromatosis. Quart Rev Biol 31:303, Review
79. Long RE (1956) A history of the therapy of tuberculosis and the case of Frederic Chopin. Quart Rev Biol 31:333–334, Review
80. McKusick VA (1956) A synopsis of clinical Auscultation, being a treatise on cardiovascular and respiratory sound, introduced by an historical survey, illustrated by sound spectrograms (spectral phonocardiograms), and supplemented by a comprehensive bibliography. privately printed and bound, in limited numbers. Baltimore
81. McKusick V, Rytand DA, Creger W (1956) Annual review of medicine, Vol 7. Quart Rev Biol 31:331–32, 31:331–332, Review
82. McKusick VA (1956) Carcinoid cardiovascular disease. Bull JHH 98:13–36
83. (1956)Fifth Macy conference on connective tissue. Bull JHH 98:62–63
84. Henry F (1956) Symposium on cardiovascular surgery. Bull JHH 98:322, Review
85. McKusick VA (1956) Heritable disorders of connective tissue, 1st edn. Mosby, St. Louis
86. McKusick VA (1956) Heritable disorders of connective tissue. IV. The Ehlers-Danlos syndrome. J Chronic Dis 3:2–24
87. McKusick VA (1956) Heritable disorders of connective tissue. V. Osteogenesis imperfecta. J Chronic Dis 3:180–023:180–202
88. McKusick VA (1956) Heritable disorders of connective tissue. VI. Pseudoxanthoma elasticum. J Chronic Dis 3:263–833:263–283
89. McKusick VA (1956) Heritable disorders of connective tissue. VII. The Hurler syndrome. J Chronic Dis 3:360–893:360–389
90. McKusick VA (1956) Heritable disorders of connective tissue. VIII. Concluding comments. J Chronic Dis 3:521–563:521–556
91. Webb GN, McKusick VA (1956) Instrumentation for spectral phonocardiography. IRE Trans Med Electronics PGME- 5:23
92. Touraine A (1956) L' L' Hérédité en Médecine. Quart Rev Biol 31:303, Review
93. Wolf S, Gordon PV Jr, Shepard EM, Wolff HG (1956) Life stress and essential hypertension. Quart Rev Biol 31:335, Review
94. McKusick VA (1956) Numerous abstracts of articles on cardiovascular topics. Circulation 13 and 14
95. Harrison TR (1956) Principles of internal medicine (1954). Quart Rev Biol 31:159–60, 31:159–160, Review
96. Franklin B (1956) Some account of the Pennsylvania hospital. Bull JHH 98:320–21, 98:320–321, Review
97. McKusick VA, Massengale ON Jr, Wigod M, Webb GN (1956) Spectral phonocardiographic studies in congenital heart disease. Br Heart J 18:403–1618:403–416
98. Kerr A Jr (1956) Subacute bacterial endocarditis. Quart Rev Biol 31:333, Review
99. Guyton AC (1956) Textbook of medical physiology, and Evans' CL principles of human physiology. Quart Rev Biol 31:316, Review
100. Myer (1956) The interpretation of the unipolar electrocardiogram and Katz' clinical electrocardiography. Bull JHH 99:47, Review
101. Montagna W (1956) The Structure and function of skin. Quart Rev Biol 31:312, Review

102. Davies GO (1956) Revision of Gaign and Davies' veterinary pathology and bacteriology. Quart Rev Biol 31:332–33, 31:332–333, Review
103. Bram WA (1956) Your blood pressure and how to live with it. Quart Rev Biol 31:334, Review
104. Webb GN, McKusick VA (1957) Analysis of heart sounds. 1957 National Conference on Instrumental Methods of Analysis. Instrument Society of Americ Proceedings, paper A157-5-4
105. Mellor RC (1957) Analytical pathology. Quart Rev Biol 32:309–10, 32:309–310, Review
106. Hall VE (1957) Annual review of physiology. Quart Rev Biol 32:299, Review
107. McKusick VA, Logue RE, Bahnson HT (1957) Association of aortic valvular disease and cystic medial necrosis of the ascending aorta: report of four instances. Circulation 16:188–9416:188–194
108. McKusick VA (1957) Symposium on cardiovascular sound. Circulation 16:270–90, 16:270–290, 414–436, 1957
109. Royal Society (1957) Biographical memoirs of fellows of the royal society vol. 1 (1952–1954) and vol. 2 (1956). Quart Rev Biol 32:281, Review
110. Rushmer RR (1957) Cardiac diagnosis: a physiologic approach. Circulation 15:772–74, 15:772–774, Review (with E.B. Burke)
111. McKusick VA, Ross RS (1957) Cardiovascular aspects of the Ehlers-Danlos syndrome, a heritable disorder of connective tissue. Circulation 16:914, Abstract
112. McKusick VA (1957) Clinical survey of cardiovascular sound. Circulation 16:424–3116: 424–431
113. Talbott JH, Ferrandi RM (1957) Collagen diseases, including systematic lupus erythematosus, polyarteritis, dermatomyositis, systematic scleroderma, thrombotic thrombocytopenic purpura. New York and London: Grune and Stratton. Quart Rev Biol 32:310, Review
114. Wood P (1957) Diseases of the heart and circulation, 2nd ed. Philadelphia: Lippincott, 1956. Bull JHH 100:190–91, 100:190–191, Review
115. Heyden KP, Sternberg ED (1957) Doctor's offices and clinics, and dental. Medical Quart Rev Biol 32:313, Review
116. Wiskine HK, McKusick VA (1957) Electric analog for analysis of hepatic hemodynamics in man. Clin Res Proc 4:18, Abstract
117. Bauer KF (1957) Ergebnisse der Medizinischen Grundlagenforschung. Quart Rev Biol 32:309, Review
118. Gardner EJ (1957) Genetics laboratory exercises. Quart Rev Biol 32:389, Review
119. McKusick VA (1957) Mechanisms in genetic diseases of man. Am J Med 22:676–8622:676–686
120. McKusick VA (1957) Numerous abstracts of articles on cardiovascular subjects. Circulation 15 and 16
121. McKusick VA (1957) Sermon delivered at Baptist Meeting-House, Parkman, Maine. 12 Sept 1832. Piscataquis Observer, Dover-Foxcroft, Me.
122. McKusick VA (1957) The functional vocabulary of valvular heart disease. Clin Res Proc 5:239
123. McKusick VA (1957) The genesis of musical murmurs. IRE Trans Med Electron 9:11–12
124. McKusick VA (1957) The genetic behavior of heritable disorders of connective tissue. Acta Genet Statist Med 7:150–547:150–154
125. Brecher GA (1957) Venous return New York: Grune & Stratton. Inc Bull JHH 101:1121, Review
126. Ross RS, McKusick VA, Harvey JC (1957) The problems of fever in patients with valvular heart disease. JAMA 165:1–7 (See also, Modern Medicine, Dec. 1, 1957, p. 164)
127. (1958) Review: Methods in Medical Research 7. Chicago: The Year Book Publishers. Quart Rev Biol 33:240–41
128. Roundtree LG (1958) Amid masters of twentieth century medicine. Bull Hist Med 31:132, Review
129. McKusick VA, Sharpe WD, Warner AO (1958) An exhibition on the history of cardiovascular sound, including the evolution of the stethoscope. Bull Hist Med 31:463–8731:463–487

130. Rytand, Creger (1958) Annual review of medicine, Vol. 7, 1956. Quart Rev Biol 33:293, Review
131. Bearn AG, McKusick VA (1958) Azure lunulae: an unusual change in the fingernails in two patients with hepatolenticular degeneration (Wilson's disease). JAMA 166:903–06166:903–906
132. Dalgaard OZ (1958) Bilateral polycystic disease of the kidneys. Enjar Munksgaards, Copenhagen, 1957. Quart Rev Biol 33:223–24, 33:223–224, Review
133. McKusick VA (1958) Cardiovascular sound in health and disease. Williams & Wilkins, Baltimore
134. McKusick VA, Fisher AM (1958) Congenital cystic diseases in the lung, with progressive pulmonary fibrosis and carcinomatosis. Ann Int Med 48:774–9048:774–790
135. Boyer SH IV, McKusick VA (1958) Disease of the aorta. Ann Rev Med 9:85–100
136. Hutt FB (1958) Genetic resistance to disease in domestic animals Ithaca: Cornell University Press 1957. Bull JHH 103:1511, Review
137. McKusick VA (1958) Genetics and public health. Virginia Med Mon 85:544–4785:544–547
138. McKusick VA (1958) Genetics in relation to cardiovascular diseases. Eugen Quart 5:30
139. Boyer SH, Chisholm AW, McKusick VA (1958) Heart disease in Friedreich's ataxia: incidence and manifestations. Circulation 18:697, Abstract
140. McKusick VA (1958) Heritable disorders of connective tissue, in Harrison's Harrison's principles of internal medicine, 3rd edn. Blakiston, New York
141. McKusick VA (1958) Numerous abstracts of articles on cardiovascular subjects. Circulation 17 and 18
142. Bohan PT (1958) Memorial lectures on medicine 1956. Quart Rev Biol 33:293, Review
143. McKusick VA (1958) Parkman, Maine: a frontier settlement. Old-Time New England 49:41
144. McKusick VA (1958) Rouanet of Paris and New Orleans; experiments on the valvular origin of the heart sounds 125 years ago. Bull Hist Med 32:137–5132:137–151
145. McKusick VA (1958) Second symposium on cardiovascular sound. Circulation 18:946–4418:946–1044
146. Roueché's Roueché's B (1958) The incurable wound. Little Brown and Co, Boston, Review
147. McKusick VA (1958) The genetic aspects of cardiovascular diseases. Ann Int Med 49:556–6749:556–567
148. Warner AO, McKusick VA (1958) Wolff-Parkinson-White syndrome a genetic study. Clin Res 6:18, Abstract
149. McKusick VA (1958) The current states of spectral phonocardiography. (Abstract) Proceedings of Third World Congress of Cardiology, Brussells, Sept 1958 p 277
150. Abel JJ (1957) A collection of papers by and about him. Williams and Wilkins, Baltimore, Am. Rev. Resp. Dis. 80: 113 only, 1959
151. McKusick VA, Talbot SA (1959) Analysis of genetic linkage in man with assistance of digital computer. Medical symposium, sponsored by IBM, Poughkeepsie, NY, June 15–17, 1959
152. (1959) Review: annual review of medicine, Vol. 10 (1959). Quart Rev Biol 34:331
153. (1959) Review: commission on chronic illnesses. Chronic illness in a large city: The baltimore study. Commonwealth Fund and Harvard University Press, 1957. Quart Rev Biol 34:183
154. McKusick VA (1959) Clinical axioms concerning heart sounds and murmurs. Curr Med Digest : 57–69
155. McKusick VA (1959) Effect of lithium on the electrocardiogram of animals. Fed Proc 9:84
156. McKusick VA, Wiskind HK (1959) Felix Savart (1791–18411, physician, physicist: early studies pertinent to the understanding of murmurs. J Hist Med 14:411–2314:411–423
157. Esterly JR, McKusick VA (1959) Genetic and physiologic studies of Milroy's disease. Clin Res 7:263, Abstract
158. McKusick VA (1959) Genetic factors in cardiovascular diseases. I. The four major types of cardiovascular disease. Mod Con Cardiovasc Dis 28:535–5128:535–551
159. McKusick VA (1959) Genetic factors in cardiovascular disease in Biostatistics and Genetics in Medical Education. Report of the 6th conference of cardiovascular training grant program directors, Hershey, Pa, 6 June 1959

160. McKusick VA (1959) Genetic factors in diseases of connective tissue: survey of the present state of knowledge. Am J Med 26:283–0226:283–302
161. McKusick VA (1959) Genetic factors of cardiovascular diseases. II. Disorders of primarily genetic etiology. Mod Concepts Cardiovasc Dis 28:547
162. McKusick VA (1959) Genetics in relation to cardiovascular diseases. In: Hammons HG (ed) Heredity counseling. Grune and Stratton, New York
163. McKusick VA (1959) Hereditary disorders of connective tissue. Bull NY Acad Med 35:143–5635:143–156
164. McKusick VA et al (1959) Medical genetics 1958. J Chronic Dis 10:256–363
165. McKusick VA, Wiskind HK (1959) Osbourne Reynolds of Manchester: contributions of an engineer to the understanding of cardiovascular sound. Bull Hist Med 33:116–3533:116–135
166. Prior JA, Silberstine JS (1959) Physical diagnosis: the history and examination of the patient. St. Louis: C.V. Mosby, 1959. Bull JHH 104:346–47, 104:346–347, Review
167. McKusick VA (1959) Spectral phonocardiography. Am J Cardiol 4:200–064:200–206
168. McKusick VA (1959) Vererbäre Störungen des Bindegewebes. (German translation by Ernst Oppenheimer.). Georg Thieme Verlag, Stuttgard
169. McKusick VA, McKusick AB (1959) Cyanotic congenital heart disease in the self portraits of an artist. Ciba Sym 7:37–40 (Also published in German, Dutch, French, Spanish and Italian.)
170. McKusick VA (1960) Trastornos hereditarios del tejido conjuntivo. America Clinica, Jan, pp Am Clinica : 32–40
171. Wood JE, Esterly JR, McKusick VA (1960) Blood flow in Milroy's disease. Clin Res 8:195, Abstract
172. Murphy DP, Abbey H (1960) Cancer in families. Bull JHH 106:58–59, Review
173. de Grouchy J, Gedda L (1960) Hérédité moleculaire. Science 131:1218, Review
174. Ferguson-Smith MA, Johnston AW (1960) Human chromosomes in disorders of sex differentiation. Trans Ass Am Phys 73:60–71, Reviews
175. McKusick VA (1960) Chromosomes in Marfan's disease. Lancet I 275:1194 (Letter)
176. McKusick VA (1960) Genetic studies in gastroenterology (symposium). LancetI 275:971–73275:971–973
177. Brinkhous, de Nicola's Nicola's (1960) Hemophilia and other hemorrhagic states. Quart J Biol 35:174
178. McKusick VA (1960) Heredity and disease of connective tissue. Ann N Y Acad Sci 86:1098–1108
179. Harri H (1960) Human biochemical genetics. Ann Int Med 52:1170–71, 52:1170–1171, Review
180. McKusick VA (1960) Inheritance of pathologic traits. McGraw Hill Encyclopedia of Science and Technology
181. de Grouchy J, Gedda L (1960) L'Hérédité L'Hérédité Moleculaire Conditions Normales et Pathologiques. Science 131:1218, Review
182. McKusick VA et al (1960) Medical genetics, 1959. J Chronic Dis 12:1202
183. McKusick VA, Young WJ, Ferguson-Smith MA (1960) Narrative for film medical genetics I. National Foundation, New York
184. McKusick VA (1960) Phonocardiography. Md State Med J 9:416
185. Steinberg (1960) Progress in medical genetics, Vol 1. Ann Int Med 55:541, Review
186. White PD, Rusk, Williams, Lee (1960) Rehabilitation of the cardiovascular patient. Ann Rev Resp Dis 82:119, Review
187. Manley KA, McKusick VA (1960) Serum cholesterol values in a hospital population. Bull JHH 107:213–27107:213–227
188. Review: Ciba Symposium Significant Trends in Medical Research. Arch. Int. Med. 106:220–21, 1960
189. McMeneey (1960) The life and time of Sir Charles Hastings. Bull JHH 106:123, Review
190. McKusick VA (1960) The genetic approach to the study of gastrointestinal disease. Gastroenterology 39:505–09, 39:505–509, Editorial

191. McKusick VA (1960) The history of methods for the diagnosis of heart disease. Bull Hist Med 34:16–18
192. McKusick VA (1960) Heritable disorders of connective tissue, 2nd edn. Mosby, St. Louis
193. McKusick VA (1960) The present status of studies of the basic defect in some heritable disorders of connective tissue. Presented at annual meeting of American Society of Human Genetics, Memphis, 29 Apr 1960
194. McKusick VA (1960) Walter S. Sutton and the physical basis of Mendelism. Bull Hist Med 34:487–497
195. McKusick VA (1960) Genetics and the nature of essential hypertension. Circulation 22:857–63, 22:857–863, Editorial
196. Evans DAP, Manley KA, McKusick VA (1960) Genetic control of isoniazid metabolism in man. Brit Med J 2:485–912:485–491
197. (1960) Ciba symposium on biochemistry of human genetics. Arch Int Med 106: 221–22 108:202, 1961
198. Eifrig DE, Imboden JB, McKusick VA, Canter AD (1961) Constrictive pericarditis: psychologic aspects of convalescence following pericardectomy. J Chronic Dis 13:52–813:52–58
199. Evans DAP, Storey PB, McKusick VA (1961) Further observations on the determination of the isoniazid inactivator phenotype. Bull JHH 108:60–6108:60–66
200. Fuller JL, McKusick VA (1961) A short course in medical genetics. Eugenics Quart 8:98
201. Fuller JL, Thompson WR (1961) Behavior genetics. Quart Rev Biol 36:218
202. McKusick VA, Harris WS (1961) The Buerger syndrome in the Orient. Bull JHH 109:241–91109:241–291
203. McKusick VA, Harris WS (1961) Buerger syndrome in the Orient. Lancet 1:1117–18, 1:1117–1118, Letter
204. McKusick VA, Harris WS (1961) The Buerger syndrome in the Orient and the United States. Circulation 24:995, Abstract
205. Skyring AP, McKusick VA (1961) Clinical genetic and electrocardiographic studies in childhood muscular dystrophy. Am J Med Sci 242:534–47242:534–547
206. Buck AA, McKusick VA (1961) Epidemiologic investigations of sarcoidosis III serum proteins; syphilis; association with tuberculosis, familial aggregation. Am J Hygiene 74:174–8874:174–188
207. Larrson T, Sjogren T (1961) Essential tremor. Quart Rev Biol 36:238
208. McKusick VA (1961) Heart sounds and the stethoscope. Curr Med Digest 28:45–52
209. Robert F (1961) Introduction to medical genetics. genetics (2nd ed). Quart Rev Biol 36:210 (2nd ed)
210. McKusick VA et al (1961) Medical genetics 1958–1960. An annotated review. Mosby, St. Louis
211. McKusick VA et al (1961) Medical genetics 1960. J Chronic Dis 14:1–98
212. Stanbury J et al (1961) The metabolic basis of inherited disease. Am Rev Resp Dis 84:755–56, 84:755–756, Review
213. Stern C (1961) Principles of human genetics. genetics (2nd ed). Bull JHH 109:52 (2nd ed.)
214. Bang FB, Wolstenholme GEW, O'Connor O'Connor CM, O'Connor O'Connor M (1961) Ciba symposium significant trends in medical research. Quart Rev Biol 36:238, Review
215. (1961) Ciba symposium significant trends in medical research. Bull JHH 108: 154
216. McKusick VA (1961) Genetic factors in cardiovascular disease. J Am Geriatr Soc 9:465–769:465–476
217. Rostand J (1961) Human heredity. Science 134:549–50, 134:549–550, Curr Med Digest 28: 36, 1961
218. McKusick VA (1962) A catalog of X-linked traits in man. J Genet Hum 11:51–64
219. Johnston AW, McKusick VA (1962) A sex-linked recessive form of spastic paraplegia. Am J Hum Genet 14:83–94
220. Niedelman ML, McKusick VA (1962) Acrokeratosis verruciformis (Hopf). A follow-up study. Arch Dermat 86:779–8286:779–782
221. (1962) Annual review of medicine (Vol. 13, 1962). Quart Rev Biol 37:274–75

222. Boyer SH IV, Chisholm AW, McKusick VA (1962) Cardiac aspects of Friedreich's ataxia. Circulation 25:493–0525:493–505
223. Boyer SH, Young WJ, McKusick VA (1962) Biochemical genetics in man. Narrative for film, medical genetics II. National Foundation, New York
224. McKusick VA, Harris WS, Ottesen OE, Goodman RM (1962) The Buerger syndrome in the United States. Arteriographic studies, with particular reference to arm involvement and the differentiation from atherosclerosis and embolism. Bull JHH 110:145–76110:145–176
225. McKusick VA, Harris WS, Ottesen OE, Goodman RM, Shelly WM, Bloodwell RD (1962) Buerger's disease: a distinct clinical and pathologic entity. Rev Surg 19:367–68, 19:367–368, Abstract
226. McKusick VA (1962) Comments. In: Blumberg BS (ed) Conference on genetic polymorphisms andgGeographic variations in disease. Grune & Stratton, New York
227. Beecher HK (1962) Disease and the advancement of basic science. Quart Rev Biol 37:65, Review
228. McKusick VA (1962) Footprints of previous pericardial disease. Jap Heart J 3:1–3, Editorial
229. McKusick VA (1962) Genetic factors in intestinal polyposis. JAMA 182:271–77182:271–277
230. Porter IH, Schulze J, McKusick VA (1962) Genetical linkage between the loci for glucose-6-phosphate dehydrogenase deficiency and colour-blindness in American Negroes. Ann Hum Genet 26:107–2226:107–122
231. Marx (1962) Health of the presidents. Bull Hist Med 36:383–84, 36:383–384, Review
232. McKusick VA (1962) Hemophilia in early New England. Bull JHH 10:177–78, 10:177–178, Abstract
233. McKusick VA (1962) Hemophilia in the early New England: a follow-up of four kindreds in which hemophilia occurred in the pre-revolutionary period. J Hist Med 42:342–6442:342–364
234. McKusick VA (1962) Heritable disorders of connective tissue. Md State Med J 11:237–4511:237–245
235. Porter IH, Schulze J, McKusick VA (1962) Linkage between glucose 6-phosphate dehydrogenase and colour-blindness. Nature 193:506
236. McKusick VA et al (1962) Medical genetics 1961. J Chronic Dis 15:417–7215:417–472
237. Stanbury JB, Wyngaarden JB, Fredricksen DS (1962) Metabolic basis of inherited disease. Bull JHH 111:322
238. McKusick VA (1962) On interviewing and being interviewed for internships. Clin Res 10:370–72, 10:370–372, Editorial
239. McKusick VA (1962) On the X chromosome of man. Quart Rev Biol 37:69–175
240. Kalow's Kalow's W (1962) Pharmacogenetics. Bull JHH 111:321–22, 111:321–322, Review
241. McKusick VA, Goodman RM (1962) Pinnal calcification: observations in systemic diseases not associated with disordered calcium metabolism. JAMA 79:230–3279:230–232
242. Smith EW, Malak JA, Goodman RM, McKusick VA (1962) Reactive perforating elastosis: a feature of certain genetic disorders. Bull JHH 111:235–51111:235–251
243. Wiggers' CJ (1962) Reminiscences and Adventures in Cardiovascular Research. Bull Hist Med 36:93, Review
244. Leeming JT, McKusick VA (1962) Serum seromucoid levels in the Marfan syndrome. Bull JHH 110:38–8110:38–48
245. Bannerman RM (1962) Thalassemia. Quart Rev Biol 37:45–637:45–46
246. Humphries JO, McKusick VA (1962) The differentiation of organic and "innocent" systolic murmurs. Prog Cardiovasc Dis 5:152–171
247. McKusick VA (1962) The earliest record of hemophilia in America. Blood 19:243–419:243–244
248. Davies SH, Gavin J, Goldsmith KLG, Graham JB, Hamper J, Hardisty RM, Harris JB, Holman CA, Ingram GIC, Jones TG, McAfee LA, McKusick VA, O'Brien O'Brien JR, Race RR, Sanger R, Tippett P (1962) The linkage relations of hemophilia A and hemophilia B (Christmas disease) to the Xg blood group system. Am J Hum Genet 15:481–9215:481–492
249. Vandenberg SG, McKusick VA, McKusick AB (1962) Twin data in support of the Lyon hypothesis. Nature 194:505–6194:505–506

250. McKusick VA (1962) Anders lecture: heritable disorders of connective tissue. Trans Stud Coll Phy Phila 29:112–2629:112–126
251. McKusick VA (1962) A form of vascular disease relatively frequent in the Orient. Am Heart J 63:57–64
252. McKusick VA, Harris WS, Ottesen OE, Goodman RM, Sheeley WM, Bloodwell RO (1962) Buerger's disease: a distinct clinical and pathologic entity. JAMA 181:5–12
253. McKusick VA, Talbot SA, Webb GN, Battersby EJ (1962) Technical aspects of the study of cardiovascular sound. In: Hamilton WF, Dow P (eds) Handbook of physiology, 1st edn, 2. American Physiologic Society, Washington, DC, pp 681–94681–694
254. McKusick VA, Rappaport SI (1962) History of classical hemophilia in a New England family. Arch Int Med 110:144–49110:144–149
255. McKusick VA (1962) On the X chromosome of man. Ann Int Med 56:991–6, 56:991–996, Editorial
256. Haws DV, McKusick VA (1963) Farabee's brachydactylous kindred revisited. Bull J H H 113:20–30
257. McKusick VA (1963) Frederick Parkes Weber: 1863–1962. JAMA 183:45–59
258. McKusick VA, Young WJ, Murphy EA (1963) Genes in families and in populations. Narrative for motion picture medical genetics III. National Foundation, New York
259. Ashley Montague MF (1963) Genetic mechanisms in human disease. Ann Rev Resp Dis 87:922, Review
260. McKusick VA, Goodman RM, Danks DM (1963) Genetic aspects of respiratory disease. Arch Environ Health 6:25–36
261. McKusick VA, Murphy EA (1963) Genetic factors in the etiology of myocardial infarction. In: James TN, Keyes JW (eds) The etiology of myocardial infarction. Little, Brown and Company, Boston, Henry Ford Hospital International Symposium
262. McKusick VA (1963) Genetics in 1962. Med World News
263. McKusick VA (1963) Genetics in medicine and medicine in genetics. Am J Med 34:594–9934:594–599
264. Goldschmidt E (1963) Genetics of migrant and isolate populations. Curr Med Digest 30:26, Review
265. McKusick VA (1963) Hemophilia in early New England. A follow-up study of the kindreds reported by Otto, Smith, Hay and the Buels (1803–1817). In: Proceedings second international congress human genetics (Rome, 1961), pp 1160–63
266. Boyd S (1963) Introduction to study of disease (5th ed.). Quart Rev Biol 38:443, Review
267. McKusick VA et al (1963) Medical genetics 1962. J Chronic Dis 16:457–634
268. McKusick VA (1963) Musical murmurs: spectral phonocardiographic studies. In: Sigal B, Likoff W (eds) The theory and practice of auscultation. F.A. Davis Co, Philadelphia, pp 93–101
269. McKusick VA (1963) Natural selection and contemporary cardiovascular disease. Circulation 27:161–63, 27:161–163, Editorial
270. Steinberg AG, Bearn AG (1963) Progress in medical genetics (Vol. 2). Quart Rev Biol 38:261, Review
271. Goodman RM, Smith EW, Paton D, Bergman RA, Ottesen OE, Siegel CL, Shelley WM, Pusch AL, McKusick VA (1963) Pseudoxanthoma elasticum: clinical and histopathologic studies. Medicine 42:297–3442:297–334
272. McKusick VA, Sanger R, Race RR et al (1963) Source of the X in XO females: the evidence of Xg. Lancet 1:558–9, 1:558–559, Letter
273. Scott PP, McKusick VA, McKusick AB (1963) The nature of osteogenesis imperfecta in cats: evidence that the disorder is primarily nutritional, not genetic, and therefore not analogous to the disease in man. J Bone Joint Surg 45A:125–3445A:125–134
274. McKusick VA (1963) Chromosomes, genes and families. In: Fishbein M (ed) Birth defects. J.B. Lippincott Co, Philadelphia
275. McKusick VA (1964) A guide to the literature in medical genetics. JAMA 188: 271–2188:271–272

276. McKusick VA (1964) Approaches and methods in human genetics. Am J Obstet Gynecol 90:1014–1023
277. McKusick VA (1964) Consanguineous marriage. Science 143:100–1, 143:100–101, Letter
278. McKusick VA, Hostetler JA, Egeland JA (1964) Genetic studies of the Amish: background and potentialities. Bull JHH 115:203–22115:203–222
279. McKusick VA, Egeland JA, Eldridge R, Krusen DE (1964) Dwarfism in the Amish. I. The Ellis-van Creveld syndrome. Bull JHH 115:307–36115:307–336
280. McKusick VA, Eldridge R, Hostetler JA, Ruangwit U, Egeland JA (1964) Dwarfism in the Amish. 11. Cartilage-hair hypoplasia. Bull JHH 116:285–26116:285–326
281. McKusick VA (1964) Dysostose metaphysaire et modification des cheveux: un nouveau (?) syndrome de transmission recessive autosomique. Presse Med 72:907–908
282. McKusick VA, Gorlin RJ, Pindborg JJ (1964) Syndrome of the head and neck. McGraw-Hill, New York
283. Milch RA, McKusick VA (1964) Genes, molecules and deformities. Clin Orthop 33:11–21
284. McKusick VA (1964) Genetic aspects of epidemiology and preventive medicine. In: Sartwell PE (ed) (Maxcy-Rosenau) preventive medicine and public health, 9th edn. Appleton-Century-Croft, New York
285. McKusick VA (1964) Genetics and the Amish. JAMA 180:850–1, 180:850–851, Editorial
286. McKusick VA (1964) Genetics in 1963. Med World News, p. 173
287. McKusick VA (1964) Human genetics. Prentice-Hall, Englewood Cliffs, N.J
288. McKusick VA (1964) Intracranial aneurysm and heredity. JAMA 190:791
289. McKusick VA (1964) Lewis A Conner Memorial Lecture: A genetical view of cardiovascular disease. Circulation 30:326–5730:326–357
290. McKusick VA (1964) Metaphyseal dysostosis with thin hair: a "new" recessively inherited syndrome? Lancet 1:832–3, 1:832–833, Letter
291. McKusick VA (1964) On the X chromosome of man. AIBS, Washington
292. Bowman HS, McKusick VA, Dronamraju KR (1964) Pyruvate kinase deficient hemolytic anemia in an Amish isolate. Am J Hum Gen 17:1–8
293. McKusick VA (1964) Some computer applications to problems in human genetics. In: Proceedings 6th IBM medical symposium, 5–9 Oct 1964
294. Milch RA, McKusick VA (1964) Introduction to the symposium on genetics and congenital deformities of the skeleton. Clin Orthop 33:1–179
295. McKusick VA, Milch RA (1964) The clinical behavior of genetic disease: selected aspects. Clin Orthop 33:22–39
296. McKusick VA et al (1964) Medical genetics 1963. J Chronic Dis 17:1077–1215
297. McKusick VA, Hostetler JA, Egeland JA, Eldridge R (1964) The distribution of certain genes in the old order Amish. Cold Spring Harb Symp Quant Biol 29:99–113
298. McKusick VA, Eldridge R, Hostetler JA, Egeland JA (1964) Dwarfism in the Amish. Trans Assoc Am Phys 77:151–68, 77:151–168, Reprinted in Birth Defects Reprint Series, National Foundation, 1965
299. McKusick VA (1964) The mucopolysaccharidoses. Ann Int Med 60:739–40, 60:739–740 Correction of author's author's names: Ann Int Med 60:940, 1964
300. McKusick VA et. al. (1964) Genetic factors in cardiovascular disease. In Andrus EC, Maxwell CH (eds) The heart and circulation. Second national conference on cardiovascular disease, vol 1. Washington, DC, (Research), pp 25–29
301. McKusick VA, Bauer RL, Koop CE, Scott RB (1964) Hydrometrocolpos: a simply inherited malformation. JAMA 189:813–16189:813–816
302. McKusick VA (1966) Heritable disorders of connective tissue. tissue, 3rd edn. Mosby, St. Louis (3rd edn.)
303. Schimke RN, McKusick VA, Huang T, Pollack AD (1965) Homocystinuria: studies of 20 families with 38 affected members. JAMA 193:711–19193:711–719
304. Hostetler JA, McKusick VA (1965) Genetic studies of the Amish: a summary and bibliography. Mennonite Quart Rev 39:223–2639:223–226
305. McKusick VA (1965) The royal hemophilia. Sci Am 213:88–95

306. Clayton RJ, Iber FL, Ruebner BH, McKusick VA (1965) Byler's disease: fetal familial intrahepaticcholestasis in an Amish kindred. J Pediatr 67:1025–8, 67:1025–1028, Abstract
307. McKusick VA (1965) Coronary artery disease. In: Need JV, Shaw MW, Schull WJ (eds) Genetics and the Epidemiology of Chronic Diseases. Public Health Serv, Washington, Pub. No. 1163
308. Gills JP, Hobson R, Hanley WB, McKusick VA (1965) Electroretinography and fundus oculi findings in Hurler's disease and allied mucopolysaccharidoses. Arch Ophthalmol 74:596–0374:596–603
309. McKusick VA (1965) Genetics in 1964. Med World News, p 141, 15 Jan 1965
310. Rimoin DL, Merimee TJ, McKusick VA (1965) Growth hormone deficiency in man: an isolated, recessively inherited defect. Science 182:1635
311. Bowen P, Lee CSN, Migeon CJ, Kaplan NM, Whalley PJ, McKusick VA, Reifenstein EC (1965) Hereditary male pseudohermaphroditism with hypogonadism, hypospadias and gynecomastia (Reifenstein's syndrome). Ann Int Med 62:252–7062:252–270
312. McKusick VA (1965) Biographies of James Dewey Watson and Maurice H.F. Wilkins. The Encyclopedia Americana
313. Goldberg MF, Maumenee AE, McKusick VA (1965) Corneal dystrophies associated with abnormalities of mucopolysaccharide metabolism. Arch Ophthalmol 74:516–2074: 516–520
314. (1965)Review: Review: Man and his future. Ciba foundation volume (1963). Bull JHH 116: 77116:77, 1965
315. McKusick VA (1965) Masks as a traffic hazard. New Eng J Med 272:487, Letter
316. McKusick VA et al (1965) Medical Genetics 1961–1963. An annotated review. Pergamon Press, Oxford
317. McKusick VA (1965) Some computer applications to problems of human genetics. Meth Inf Med 4:183–94:183–189
318. McKusick VA (1965) The genetic mucopolysaccharidoses. Circulation 31:1–4, Editorial
319. McKusick VA (1965) The role of heredity in diseases of the gastrointestinal system. In: Paulson M (ed) Gastroenterologic diseases. Lea and Febiger, Philadelphia
320. McKusick VA, Schimke RN, Huang T, Pollack AD (1965) Thrombotic vascular disease in homocystinuria, a newly recognized inborn error of metabolism simulating Marfan's syndrome. Circulation 32(Suppl):149–50, 32(Suppl):149–150, Abstract
321. Blyth H, Carter CO, Dubowitz V, Emery AEH, Gavin J, Johnston HA, McKusick VA, Race RR, Sanger R, Tippett P (1965) Duchenne's muscular dystrophy and the Xg blood groups: a search for linkage. J Med Genet 2:157–602:157–160
322. McKusick VA (1965) Generalized genetic disorders of the osseous skeleton. JAMA 191:754–5, 191:754–755, Editorial
323. Schimke RN, McKusick VA, Pollack AD (1965) Homocystinuria simulating Marfan's syndrome. Trans Assoc Amer Phys 78:60–278:60–62
324. Strasburger AK, Hawkins MR, Eldridge R, Hargrave RL, McKusick VA (1965) Symphalangism: genetic and clinical aspects. Bull JHH 117:108–27117:108–127
325. McKusick VA, Kaplan D, Wise D, Hanley WB, Suddarth SB, Sevick MD, Maumenee AE (1965) The genetic mucopolysaccharidoses. Medicine 44:445–9344:445–493
326. McKusick VA, McKusick VA, McKusick VA, McKusick VA (1965) Genetic factors in arteriosclerosis, with particular reference to atherosclerosos of the coronary arteries. In: Blumenthal H, Cowdry E (eds) Cowdry's Arteriosclerosis. Charles C Thomas, Springfield, Ill, Chapter 27 of part VI
327. McKusick VA (1966) Clinical genetics at a populational level: the ethnicity of disease in the United States. Ala J Med Sci 3:408–243:408–424
328. McKusick VA (1966) A concise outline of medical genetics or what the practitioner should know about medical genetics. Med Times 94:807
329. McKusick VA (1966) Heritable disorders of connective tissue: newer aspects. Structural organization of the skeleton symposium, Johns Hopkins Hospital. Birth Defects 2(1) April, 3 Nov 1965

330. McKusick VA, Stauffer M, Knox DL, Clark DB (1966) Chorioretinopathy with hereditary microcephaly. Arch Ophthalmol 75:587
331. McKusick VA, Pollack AD (1966) Clinicopathologic conference. Bull JHH 119:150–60119: 150–160
332. Konigsmark B, McKusick VA (1966) Hereditary deafness. Volta Rev 68:330–334
333. McKusick VA (1966) Hereditary disorders. Medical opinion and review I: No. 10.
334. McKusick VA (1966) Homocystinuria (Letter). Arch Ophthalmol 75:150
335. McKusick VA (1966) Human genetics: inheritance of pathologic traits. Encyclopedia of science and technology. McGraw-Hill
336. Kwiterovich PO, Cross HE, McKusick VA (1966) Mongolism in an inbred population. Bull JHH 119:268–5119:268–275
337. Rimoin D, Merimee T, McKusick VA (1966) Sexual ateliosis: a recessively inherited isolated deficiency of pituitary growth hormone. Trans Assoc Am Phys 79:297–311
338. Rimoin DL, Merimee TJ, McKusick VA (1966) Sexual ateliotic dwarfism: a recessively inherited isolated deficiency of human growth hormone. In: (Abstract) proceedings third internatonal congress of human genetics
339. McKusick VA, Cross HE (1966) Ataxia-telangiectasia and Swiss-type agammaglobulinemia. JAMA 195:739–5195:739–745
340. McKusick VA, Bias WB, Cross HE (1966) Blood groups in two old order Amish demes. In: (Abstract) Proceedings third international congress of human genetics, p 24
341. McKusick VA (1966) Mendelian inheritance in man. Catalogs of Autosomal Dominant, Autosomal Recessive and X-linked Phenotypes, 1st edn. Johns Hopkins University Press, Baltimore
342. McKusick VA (1967) Genetics to the modern era. Medical opinion and review 3: No. 3
343. Rimoin DL, Merimee TJ, Rabinowitz D, McKusick VA, Cavalli-Sforza LL (1967) Growth hormone in African pygmies. Lancet 2:523–62:523–526
344. Mahloudji M, Teasdall RD, Adamkiewicz JJ, McKusick VA (1967) Hereditary amyloid neuropathy. Trans Amer Neurol Assoc 92:46–992:46–49
345. Mahloudji M, McKusick VA (1967) Hereditary amyloidosis, Indiana type. Ann Intern Med 60:1029–130, 60:1029–1130, Abstract
346. McKusick VA (1967) Identification of abdominal murmurs. New Eng J Med 276:1207, Editorial
347. Aronson SM, Herzog ME, Brunt PW, McKusick VA, Myrianthopoulos NE (1967) Inherited neurologic diseases of Ashkenazic Jewry: demographic data suggesting nonrandom gene frequencies. Trans Amer Neurol Assoc 92:117–121
348. McKusick VA (1967) Men after mendel. Medical opinion and review 3: No. 2
349. Harvey AM, McKusick VA (1967) Osler's Osler's testbook revisited: a reprint of selected sections with commentaries. Appleton-Century-Crofts, New York
350. Merimee TJ, Rabinowitz D, Riggs L, Burgess JA, Rimoin DL, McKusick VA (1967) Plasma growth hormone after arginine infusion: clinical studies. New Eng J Med 276:434–39276: 434–439
351. McKusick VA (1967) Projection for medical genetics. Medical opinion and review 3: No. 4
352. Mengel MC, Konigsmark BW, Berlin CI, McKusick VA (1967) Recessive early-onset neural deafness. Acta Laryngol 64:313–2664:313–326
353. McKusick VA, Norum RA, Farkas HJ, Brunt PW, Mahloudji M (1967) The Riley-Day syndrome–observations on genetics and survivorship. Israel J Med Sci 3:372–93:372–379
354. McKusick VA (1967) In: Caceres CA, Perry LW (eds) Contribution to the innocent murmur: a problem in clinical practice, Little, Brown & Co., Boston
355. McKusick VA (1967) The ethnic distribution of disease in the United States. J Chronic Dis 20:115–1820:115–118
356. Schimke RN, McKusick VA, Weilbaecher RG (1967) Homocystinuria. In: Nyham WL (ed) Amino acid metabolism and genetic variation. McGraw-Hill, New York, pp 297–314
357. Niedermeyer E, McKusick VA, Brunt PW, Mahloudji M (1967) The EEG in familial dysautonomia (Riley-Day syndrome). Electroenceph Clin Neurophysiol 22:473–522:473–475

358. Cross HE, McKusick VA (1967) The Mast syndrome: a recessively inherited form of presenile dementia with motor disturbances. Arch Neurol 16:1–13
359. Cross HE, McKusick VA, Breen W (1967) A new oculocerebral syndrome with hypopigmentation. J Pediat 70:398–0670:398–406
360. Cross HE, McKusick VA (1967) The Troyer syndrome: a recessive form of spastic paraplegia with distal muscle wasting. Arch Neurol 16:473–8516:473–485
361. Hanley WB, McKusick VA, Barranco FT (1967) Osteochondritis dissecans with associated malformation in two brothers. J Bone Joint Surg 49-A:925–3749-A:925–937
362. Cross HE, McKusick VA (1967) A survey of neurological disorders in a genetic isolate. Neurology 17:743
363. McKusick VA, Mahloudji M, Abbott MH, Lindenbert R, Kepas D (1967) Seckel's birdheaded dwarfism. New Eng J Med 277:279–85277:279–285
364. Cross HE, Kennel EE, Lilienfeld AM, McKusick VA (1967) Cancer of the cervix in the Amish. Trans Assoc Amer Phys 80:133–141
365. McKusick VA (1967) Clinical genetics at a populational level: the ethnicity of disease in the United States. J Med Sci 3:408–424
366. McKusick VA, Mutalik GS (1968) Genetics and pulmonary disease. In: Liebow AA, Smith DE (eds) The lung. Williams and Wilkins, Baltimore, p 187
367. Merimee TJ, Hall JG, Rabinowitz D, McKusick VA, Rimoin DL (1968) An unusual variety of endocrine dwarfism: subresponsiveness to growth hormone in a sexually mature dwarf. Lancet 2:191–32:191–193
368. McKusick VA, Clayton RJ (1968) Cholestasis of pregnancy. New Eng J Med 278:566–7, 278:566–567, Letter
369. Eldridge R, Berlin CI, Money JW, McKusick VA (1968) Cochlear deafness, myopia and intellectual impairment in an Amish family. Arch Orolaryng 88:49–54
370. McKusick VA (1968) Genetics in the practice of medicine. Ann Roy Coll Phys Surg Canad 1:210–151:210–215
371. Rimoin DL, Merimee TJ, Rabinowitz D, McKusick VA, Cavalli-Sforza LL (1968) Growth hormone in African pygmies. Lancet 1:596, Letter
372. Rubin A (1968) Handbook of congenital malformations. JAMA 204:935–6, 204:935–936, Review
373. McKusick VA (1968) Hereditary disturbances of the connective tissues: current status. Medicina (Buenos Aires) 28:387–1228:387–412
374. Merimee TJ, Rabinowitz D, Hall J, Rimoin D, McKusick VA (1968) Isolated growth hormone deficiency: IV. The response of sexual ateliotic dwarfs to exogenous growth hormone. Metabolism 17:1012–1817:1012–1018
375. Merimee TJ, Rabinowitz D, Rimoin DL, McKusick VA (1968) Isolated hormone deficiency. III. Insulin secretion in sexual ateliotic dwarfism. Metabolism 17:1005–1117:1005–1011
376. Donahue RP, Renwick JH, Delos Cobos L, Borgaonkar DS, Bias WB, McKusick VA (1968) Karyotypic and linkage analyses in two pedigrees with marker chromosomes. Clin Res 16:296, Abstract
377. Kaplan D, McKusick VA, Trebach S, Lazarus R (1968) Keratousulfate-chondroitin sulfate peptide from normal urine and from urine of patients with Morquio syndrome (mucopolysaccharidoses IV). J Lab Clin Med 71:48–55
378. Merimee TJ, Rimoin DL, Cavalli-Sforza L, Rabinowitz D, McKusick VA (1968) Metabolic effects of human growth hormone in the African pygmy. Lancet 2:194–52:194–195
379. McKusick VA (1968) Mendelian inheritance in man. Catalogs of autosomal dominant, autosomal recessive and X-linked phenotypes, 2nd edn. Johns Hopkins University Press, Baltimore
380. Merimee TJ, Rimoin DL, Rabinowitz D, Cavalli-Sforza L, McKusick VA (1968) Metabolic studies in the African pygmy. Trans Assoc Am Phys 81:221–3081:221–230
381. McKusick VA (1968) Milroy's disease. JAMA 204:174, Letter
382. Donahue RP, Bias WB, Renwick JH, McKusick VA (1968) Probable assignment of the Duffy blood group locus to chromosome I in man. Proc Nat Acad Sci 61:949–5561:949–955

383. Review: rare and radical–abnormal hemoglobins in human populations. Med Opin Rev 4:59 only
384. Rimoin DL, Fletcher BD, McKusick VA (1968) Spondylocostal dysplasia, a dominantly inherited form of short trunk dwarfism. Am J Med 45:948–5345:948–953
385. Borgaonkar DS, Murdoch JL, McKusick VA, Borkowf SP, Money JW, Robinson WB (1968) The YY syndrome. Lancet 2:461–22:461–462
386. McKusick VA (1968) Genetics disorders involving growth hormone in man. Excerpta Medica International Congress Series. In: Proceedings of third International Congress of Endocrinology, Mexico pp 338–9
387. McKusick VA (1968) The ethnicity of disease in the United States. In: Stevenson LG, Multhauf RP (eds) Medicine, science and culture. Johns Hopkins Press, Baltimore, pp 185–207
388. Cross HE, Hollander CS, Rimoin DL, McKusick VA (1968) Familial athyreotic cretinism accompanied by muscular hypertrophy. Pediatrics 41:413–2041:413–420
389. Rimoin DL, Holzman GB, Merimee TJ, Rabinowitz D, Barnes AC, Tyson JEA, McKusick VA (1968) Lactation in the absence of human growth hormone. J Clin Endocrinol Metab 28:1183–1828:1183–1188
390. McKusick VA, Weilbaecher RG, Gragg GW (1968) Recessive inheritance of a congenital malformation syndrome: unilateral absence deformity of leg and congenital cataracts. JAMA 204:113–8204:113–118
391. Cross HE, Lerberg DB, McKusick VA (1968) Type II syndactyly. Am J Hum Genet 20:368–8020:368–380
392. McKusick VA, Cross HE (1968) Genealogical linkage of records in two isolate populations. In: Record Linkage in Medicine (proceedings of the international symposium July 1967), E. & S. Livingstone Ltd., pp 264–70
393. McKusick VA (1968) Some principles of medical genetics. In: Mihaly B (ed) Genetics in medical practice. J.B. Lippincott Co, Philadelphia, pp 43–54
394. Rimoin DL, Merimee TJ, Rabinowitz D, Cavalli-Sforza LL, McKusick VA (1968) Genetic aspects of isolated growth hormone deficiency. In: Growth Hormone, (proceedings of the international symposium 11–13 Sept, Milan, Italy), pp 418–32
395. Bloor CM, McKusick VA (1968) The genetics of coronary artery disease. In: Symposium on coronary heart disease. American Heart Association Monograph No. 2, New York, pp 6–11
396. Rimoin DL, Merimee TJ, Rabinowitz D, McKusick VA (1968) Genetic aspects of clinical endocrinology. Recent Prog Horm Res 24:365–437
397. Mahloudji M, Teasdall RD, Adamkiewicz JJ, Hartmann WH, Lambird PA, McKusick VA (1969) The genetic amyloidoses with particular reference to hereditary neuropathic amyloidosis, type II (Indiana of Rukavina type). Medicine 48:1137
398. McKusick VA (1969) Human genetics, 2nd edn. Prentice-Hall, Englewood Cliffs, NJ
399. McKusick VA (1969–1972) I. Special lectures. II. Malformation Syndromes. III. Limb Anomalies. IV. Nervous System. VII. Muscle. VIII. Eye. IX. Ear. X. The endocrine system. XI. Orofacial structures. XII. sSin, hair and nails. XIII. G.F. Tract including liver and pancreas. In: Bergsma D (ed) Clinical delineation of birth defects. Williams and Wilkins, Baltimore
400. Merimee TJ, Hall JG, Rimoin DL, McKusick VA (1969) A metabolic and hormonal basis for classifying ateliotic dwarfs. Lancet 1:963–51:963–965
401. Murdoch JL, Walker BA, Hall JG, Abbey H, Smith KK, McKusick VA (1969) Achondroplasia–a genetic and statistical survey. Ann Hum Genet 33:227–244
402. Clayton RJ, Iber FL, Ruebner BH, McKusick VA (1969) Byler disease: fatal familial intrahepatic cholestasis in an Amish kindred. Am J Dis Child 117:112–24117:112–124
403. Mengel MC, Konigsmark BW, Berlin CI, McKusick VA (1969) Conductive hearing loss and malformed low-set ears, as a possible recessive syndrome. J Med Genet 6:14–21
404. Borgaonkar DS, McKusick VA, Herr ML, de los Cobos L, Yoder OC (1969) Constancy of length of human Y chromosome. Ann Genet 12:262–264
405. McKusick VA (1969) Family oriented follow-up. J Chronic Dis 2:1–7

406. McKusick VA (1969) Fatal varicella. JAMA 207:370, Letter
407. Schleutermann DA, Bias WB, Murdoch JL, McKusick VA (1969) Linkage of the loci for the nail-patella syndrome and adenylate kinase. Am J Hum Genet 21:606–3021:606–630
408. McKusick VA (1969) Lumpers and splitters, or the nosology of genetic disease. Birth Defects 5:23–32
409. McKusick VA (1969) Medical genetics: the unifying subject. Hosp Pract 4(10):9, Editorial
410. McKusick VA (1969) On lumpers and splitters, or the nosology of genetic disease. Perspect Biol Med 12:298–312
411. Rimoin DL, Merimee TJ, Rabinowitz D, Cavalli-Sforza L, McKusick VA (1969) Peripheral subresponsiveness to human growth hormone in African pygmies. New Eng J Med 281:1383–8281:1383–1388
412. McKusick VA (1969) Population genetics, in Medicine in the University and Community of the Future. In: Purkis IE, Matthews UF (eds) Proceedings of the scientific sessions marking the centennial of the faculty of medicine, Dalhousie University, Halifax, NS 11–13 Sept 1968
413. Sluiter-Eringa H, Sluiter HJ (1969) Pulmonary arteriovenous fistula: a case-finding study and clinico-laboratory analysis of 27 cases. JAMA 210:2286, Review
414. Rimoin DL, McKusick VA (1969) Somatic mosaicism in an achondroplastic dwarf. Birth Defects 5(4):17–19
415. McKusick VA (1969) Some principles of medical genetics. Matern Child Health 1:4–5
416. Bias WB, Light-Orr JK, Krevans JR, Humphrey RL, Mamill PVV, Cohen BH, McKusick VA (1969) The Stoltzfus blood groups: a new polymorphism in man. Am J Hum Genet 21:552–821:552–558
417. Temtamy S, McKusick VA (1969) Synopsis of hand malformations with particular emphasis on genetic factors. Birth Defects 5:125–845:125–184
418. McKusick VA (1969) The role of heredity in diseases of the gastrointestinal system. In: Paulson M (ed) Gastroenterologic medicine. Lea & Febiger, Philadelphia
419. Hall JG, Levin J, Kuhn JP, Ottenheimer EJ, van Berkum P, McKusick VA (1969) Thrombocytopenia with absent radius. Medicine 48:411–3948:411–439
420. McKusick VA (1969) Treatment and prevention in genetic diseases. Matern Child Health 2:38–92:38–39
421. Mengel MC, Konigsmark BW, McKusick VA (1969) Two types of congenital recessive deafness. Eye Ear Nose Throat Mon 48:301–305
422. Hall JG, Dorst JP, Taybi H, Scott CI, Langer LO, McKusick VA (1969) Two probable cases of homozygosity for the achondroplasia gene. Birth Defects 5(4):24–34
423. McKusick VA (1969) The nosology of the mucopolysaccharidoses. Am J Med 47:730–747
424. Schleutermann DA, Bias WB, McKusick VA (1970) A kindred of koilonychia: linkage data. Am J Hum Genet 22:390–395
425. Harris H, Hirschhorn K (1970) Advances in human genetics. Science 170:1296, Review
426. Cross HE, McKusick VA (1970) Amish demography. Soc Biol 17:83–101
427. Steinhauser RK, Scott CI, Mays JA, McKusick VA (1970) Ankylosing spondylosis in familial hypophosphatemia. In: (Abstract) American College of Physicians 51st Annual Session, Philadelphia, 13 Apr 1970
428. Filippi G, McKusick VA (1970) The Beckwith-Wiedemann syndrome (the exomphalosmacroglossia- gigantism syndrome). Report of two cases and review of the literature. Medicine 49:279–298
429. Lux SE Jr, Johnston RB, August CS, Say B, Penschaszadeh VB, Rosen FS, McKusick VA (1970) Chronic neutropenia and abnormal cellular immunity in cartilage-hair hypoplasia. New Eng J Med 282:321–236
430. Mahloudji M, Brunt PW, McKusick VA (1970) Clinical neurological aspects of familial dysautonomia. J Neurol Sci 11:383
431. Fraser FC, McKusick VA (eds): Congenital Malformations (Proceedings of the third international conference, The Hague, Netherlands, 7–13 Sept 1969), Amsterdam-New York: Excerpta Medica

432. Merimee TJ, Fineberg SE, McKusick VA, Hall J (1970) Diabetes mellitus and sexual ateliotic dwarfism: a comparative study. J Clin Invest 49:1096–1102
433. McKusick VA (1970) Dietogenetics. JAMA 214:2058, Letter
434. Reed WB, Wuepper KD, Epstein JH, Redeker A, Simpson RJ, McKusick VA (1970) Erythropoietic protoporphyria. JAMA 214:1060–1066
435. Brunt PW, McKusick VA (1970) Familial dysautonomia: a report of genetic and clinical studies, with a review of the literature. Medicine 49:343–374
436. Goldberg MF, Scott CI, McKusick VA (1970) Hydrocephalus and papilledema in the Maroteaux-Lamy syndrome (mucopolysaccharidosis type VI). Am J Ophthalmol 69:969–976
437. Tyson JE, Barnes AC, Merimee TJ, McKusick VA (1970) Isolated growth hormone deficiency: studies in pregnancy. J Clin Endocrinol 31:147–152
438. McKusick VA (1970) Metachromasia elaborated. New Eng J Med 282:102–103, Letter
439. Spranger JW, Koch F, McKusick VA, Natzschka J, Wiedemann HR, Zellweger H (1970) Mucopolysaccharidosis VI (Maroteaux-Lamy's disease). Helv Paediat Acta 25:337–362
440. Tyson JE, Barnes AC, McKusick VA, Scott CI, Jones GS (1970) Obstetric and gynecologic considerations of dwarfism. Am J Obstet Gynecol 108:688–704
441. Ian Shine's Shine's Serendipity in St. Helena (1970) A clinical and medical study of an isolated community. Pergamon press, 1970. JAMA 214: 1330–1331
442. Murdoch JL, Walker BA, Hall JG, Abbey H, Smith KK, McKusick VA (1970) Achondroplasia–a genetic and statistical survey. Ann Hum Genet 33:227–245
443. Andrade C, Araki S, Block WD, Cohen AS, Jackson CE, Kuroiwa Y, McKusick VA, Nissim J, Sohar E, Van Allen MW (1970) Hereditary amyloidosis. Arth Rheum 13:902–915
444. McKusick VA (1970) The relative frequency of the Hurler and Hunter syndromes. New Eng J Med 283:853–854
445. McKusick VA (1970) The relative frequency of the Hurler and Hunter syndromes. New Eng J Med 283:1467–1468, Letter
446. McKusick VA (1970) Birth defects–Prospects for progress. In: Fraser FC, McKusick VA (eds) Congenital malformations. Excerpta Medica, Amsterdam-New York
447. Bulmer's Bulmer's MG (1970) The biology of twinning in man. Science 170:965, Review
448. McKusick VA (1970) Connective tissue. In: Goodman RM (ed) Genetic disorders of man. Little, Brown and Co, Boston, pp 331–357
449. McKusick VA (1970) Genetics, in Britannica yearbook 1970, pp 499–500
450. McKusick VA (1970) Genetic variation in man. In: Sober HA, Harte RA (eds) Handbook of biochemistry: selected data for molecular biology, 2nd edn. The Chemical Rubber Co, Cleveland, Ohio, pp 1–114, 11–118
451. Cross HE, McKusick VA (1970) Amish demography. Soc Biol 17:83–101
452. Merimee TJ, Siperstein MD, Fineberg SE, McKusick VA (1970) The microangiopathic lesions of diabetes mellitus: an evaluation of possible causative factors. Trans Ass Am Physic 83:102–112
453. Pierce ER, Weisbord T, McKusick VA (1970) Gardner's syndrome: formal genetics and statistical analysis of a large Canadian kindred. Clin Genet 1:65–80
454. McKusick VA (1970) Human genetics. In: Roman HL (ed) Annual review of genetics, vol 4. Ann. Reviews Inc, California, pp 1–46
455. McKusick VA, Bartter FC (1971) Dedication to the memory of two giants of endocrinology. Birth Defects 7(6):1–4
456. McKusick VA (1971) A mental retardation syndrome with peripheral dysotosis and pug nose. Birth Defects 7(1):249–250
457. McKusick VA, Scott CI (1971) A nomenclature for constitutional disorders of bone. J Bone Joint Surg 53-A:978–986
458. Halpern BL, Char F, Murdoch JL, Horton WB, McKusick VA (1971) A prospectus on the prevention of aortic rupture in the Marfan syndrome with data on survivorship without treatment. Johns Hopkins Med J 129:123–129
459. Harper P, Bias WB, Hutchinson JR, McKusick VA (1971) ABH secretor status of the fetus: a genetic marker identifiable by amniocentesis. J Med Genet 8:438–440

460. Stevenson RE, Howell RR, McKusick VA (1971) Corneal clouding in mucopolysaccharidosis VI (Maroteaux-Lamy syndorme). Birth Defects 7(3):156–158
461. McKusick VA (1971) The third conference on the clinical delineation of birth defects, orofacial sructures. Dedication to the memory of Jonathan Hutchinson (1828–1913). Birth Defects 7(8):1–4
462. McKusick VA (1971) Foreword to clinical diagnostic pearls. In: Birnholz JC, Michelson PE (eds) New York Medical Examination Publishing Co. Inc.,
463. McKusick VA (1971) Fourteen genetic misconceptions. Ann Int Med 75:642–643, Editorial
464. Borgaonkar DS, McKusick VA (1971) Human chromosome fluorescence. Johns Hopkins Med J 128:75–78
465. Walker BA, Murdoch JL, McKusick VA, Langer LO, Beals RK (1971) Hypochondroplasia. Am J Dis Child 122:95–104
466. McKusick VA (1971) Introduction: aims and themes of the second conference on the clinical delineation of birth defects. Birth Defects 7(1):2–4
467. McKusick VA (1971) Mendelian inheritance in man. Catalogs of Autosomal Dominant, Autosomal Recessive and X-linked Phenotypes, 3rd edn. Johns Hopkins University Press, Baltimore
468. Grewe J, Jorgenson R, McKusick VA (1971) Model system of craniofacial growth: Marfan's syndrome. J Dent Res 50:1501–1502, Abstract
469. Nickel JR, McKusick VA (1971) Nemaline myopathy: second autopsied case in a previously reported family. Birth Defects 7(2):128–129
470. Goldberg MF, McKusick VA (1971) The second conference on the clinical delineation of birth defects, eye. Dedication to Frank Burton Walsh. Birth Defects 7(3):1–2
471. Young LW, Radebaugh JF, Rubin P, Sensenbrenner JA, Fiorelli G, McKusick VA (1971) New syndrome manifested by mandibular hypoplasia, acroosteolysis, stiff joints and cutaneous atrophy (mandibuloacral dysplasia) in two unrelated boys. Birth Defects 7(7):291–297
472. McKusick VA (1971) Pachyonychia congenita in father and son. Birth Defects 7(8):274–275
473. McKusick VA (1971) Refsum's disease (heredopathia atactica polyneuriformis). Birth Defects 7(1):238
474. Esterly NB, McKusick VA (1971) Stiff skin syndrome. Pediatrics 47:360–369
475. McKusick VA (1971) Symphalangism and deafness. Birth Defects 7(4):124
476. Jones HW, McKusick VA, Harper PS, Wuu KD (1971) The HeLa cell and reappraisal of its origin. Obstet Gyncecol 38:945–949
477. McKusick VA (1971) The mapping of human chromosomes. Sci Am 224:104–113
478. McKusick VA (1971) Transverse vaginal septum (hydrometrocolpos). Birth Defects 7(6):326–327
479. Goldberg MF, McKusick VA (1971) X-linked colobomatous microphthalmos and other congenital anomalies: a disorder resembling Lenz's dysmorphogenetic syndrome. Am J Ophthalmol 71:1128–111371:1128–1133
480. McKusick VA (1971) Fourteen genetic misconceptions. Ann Int Med 75:642–645
481. Robinow M, Pfeiffer RA, Gorlin RJ, McKusick VA, Renuart AW, Johnson GF, Summitt RL (1971) Acrodysostosis: a syndrome of peripheral dysostosis, nasal hypoplasia and mental retardation. Am J Dis Child 121:195–203
482. McKusick VA, Tumulty PA (1971) The second conference on the clinical delineation of birth defects. Dedication to Frank R. Ford, M.D. Birth Defects 7(1):1
483. Halpern BL, Char FL, Murdoch JL, Horton WB, McKusick VA (1971) A prospectus on the prevention of aortic rupture in the Marfan syndrome with data on survivorship without treatment. Johns Hopkins Med J 129:123–129
484. McKusick VA, Hall JG, Char F (1971) The clinical and genetic characteristics of homocystinuria. In: Carson NAJ, Raine DN (eds) Inherited disorders of sulphur metabolism. Churchill Livingstone, London, pp 179–203
485. McKusick VA (1971) Transverse striae distensae in the lumbar area in a father and two sons. Birth Defects 7(8):260–261
486. McKusick VA (1971) Chauveau, Jean-Baptiste Auguste. Dictionary Sci Biog III:219–220

487. McKusick VA (1972) Heritable disorders of connective tissue, 4th edn. Mosby, St. Louis
488. McKusick VA (1972) Association of congenital bicuspid aortic valve and Erdheim's cystic medical necrosis. Lancet I:1026–1027, Letter
489. McKusick VA (1971) The third conference on the clinical delineation of birth defects, skin, hair and nails. Dedication to the memory of Jonathan Hutchinson. Birth Defects 7(8):1–4
490. James AE, Dorst JP, Mathews ES, McKusick VA (1972) Hydrocephalus in achondroplasia: studies by cisternography. Pediatrics 49:46–49
491. McKusick VA, Howell RR, Hussels IE, Neufeld EF, Stevenson RE (1972) Allelism, non-allelism and genetic compounds among the mucopolysaccharidoses: corrective factors in nosology, genetics and therapy. Trans Assoc Amer Phys 85:151–171
492. Chase GA, McKusick VA (1972) Controversy in human genetics: founder effect in Tay-Sachs disease. Am J Hum Genet 24:339–340
493. McKusick VA, David Y-Y (1972) Hsia–In memoriam. Clin Genet 3:151, Obituary
494. Harper PS, Rivas ML, Bias WB, Hutchinson JR, Dyken PR, McKusick VA (1972) Genetic linkage confirmed between the locus for myotonic dystrophy and the ABH-secretion and Lutheran blood group loci. Am J Hum Genet 24:310–316
495. Penchaszdeh VB, Hardy JB, Mellits ED, Cohen BH, McKusick VA (1972) Growth and development in an "inner city" poplation: an assessment of possible biological and environmental influences. I. Intra-uterine growth. Johns Hopkins Med J 130:384–397
496. Penchaszadeh VB, Hardy JB, Mellitis ED, Cohen BH, McKusick VA (1972) Growth and development in an "inner city" population: an assessment of possible biological and environmental influences. II. The effect of certain maternal characteristics on birth weight, gestational age and intra-uterine growth. Johns Hopkins Med J 131:11–23
497. McKusick VA (1972) Human chromosomes in a new light. (Editorial) Hosp Pract
498. McKusick VA et al (1972) Medical genetics, in contributions of the biological sciences to human welfare. Fed Proc 31:TF56–TF57
499. Traisman HS, Bianchine JW, McKusick VA (1972) More speculation on Marfan syndrome. J Pediat 80:530–531, Letter
500. Bias WB, Amos DB, Ward FE, Yoder OC, Renwick JH, McKusick VA (1972) No linkage between HL-A and haptoglobin loci. Am J Hum Genet 24:354–355, Letter
501. Murdoch JL, Walker BA, McKusick VA (1972) Parental age effects on the occurrence of new mutations for the Marfan syndrome. Ann Hum Genet 35:331–336
502. Steinberg AG, Bearned AG (1972) Progress in Medical Genetics. Quart Rev Biol VIII
503. Lichtenstein J, Warson R, Jorgenson R, Dorst JP, McKusick VA (1972) The tricho-dento-osseous (TDO) syndrome. Am J Hum Genet 24:559–582
504. Murdoch JL, Walker BA, Halpern BL, Kuzma JW, McKusick VA (1972) Life expectancy and causes of death in the Marfan syndrome. New Eng J Med 286:804–808
505. Murphy EA, McKusick VA (1972) Genetica del enanismo. In: Labor SA (ed) Genetica al Dia VII. Sindromes polimalformativos, Barcelona
506. Lie SO, McKusick VA, Neufeld EF (1972) Simulation of genetic mucopolysaccharidoses in normal human fibroblasts by alternation of pH of the medium. Proc Nat Acad Sci 69:2361–2363
507. Walker BA, Scott CI, Hall JG, Murdoch JL, McKusick VA (1972) Diastrophic dwarfism. Medicine 51:41–59
508. McKusick VA, Howell RR, Hussels IE, Neufeld EF, Stevenson RE (1972) Allelism, non-allelism and genetic compounds among the mucopolysaccharidoses. Lancet 1:993–996
509. McKusick VA (1972) The Bar Harbor course in medical genetics. Science 176:820–821
510. Lichtenstein JR, Bilbrey GL, McKusick VA (1972) Clinical and probably genetic heterogeneity within mucopolysaccharidosis II. Report of a family with a mild form. Johns Hopkins Med J 131:425–435
511. Melhem R, Dorst JP, Scott CI Jr, McKusick VA (1973) Roentgen findings in mucolipidosis III (Pseudo-Hurler Polydystrophy). Radiology 106:153–160
512. Kopits SE, Perovic MN, McKusick VA, Robinson RA, Bailey JA (1972) Congenital atlanto-axial dislocations in various forms of dwarfism. In: (Abstract) 39th Annual meeting American acadamy orthopaedic surgeons 29 Jan–3 Feb 1972, Washington DC

513. McKusick VA (1972) Study guide, human genetics, 2nd edn. Prentice Hall, Englewood Cliffs, NJ
514. McKusick VA, Martin GE, Lichtenstein JR et al (1973) Acquired and heritable defects in collagen synthesis and fibrogenesis. Trans Am Clin Climatol Assoc 85:130–141
515. McKusick VA (1973) Max Brodel. Dictionary Am Biog 3:106–107 (June 8, 1870-Oct. 26, 1941)
516. Rimoin DL, Hollister DW, Siggers D, Silberberg R, Lachman R, McAlister W, Kaufman R, McKusick VA, Dorst J (1973) Clinical, radiographic, histologic and ultratructural definition of the Kniest syndrome. Society for Pediatric Research, Atlantic City, Abstract
517. McKusick VA (1973) Congenital deafness and Hirschsprung's disease. New Eng J Med 288:691, Letter
518. Lindstrom JA, Bias WB, Schimke RN, Rivas ML, Chase GA, McKusick VA (1973) Genetic linkage in Huntington's disease. Adv Neurol I:203–308
519. McKusick VA (1973) Genetic studies in American inbred populations with particular reference to the old order Amish. Israel J Med Sci 9:1276–1284
520. McKusick VA (1973) Genetically determined disorders of connective tissue. Medcom, Inc, New York, Audio-visual teaching program
521. McKusick VA, Chase GA (1973) Human genetics. Ann Rev Genet 7:435–472
522. McKusick VA (1973) Mapping the chromosomes of man. In: Proceedingsof Wilson Day JC, University of Rochester, 1 Nov 1973
523. McKusick VA, Claiborne R (eds) (1973) Medical genetics. Hospital Practice Publishing Co. Inc., New York
524. McKusick VA, Kelly TE, Dorst JP (1973) Observations suggesting allelism of the achondroplasia and hypochondroplasia genes. J Med Genet 10:11–16
525. Lichtenstein JR, Kohn LD, Martin GR, Byers P, McKusick VA (1973) Procollagen peptidase deficiency in a form of the Ehlers-Danlos syndrome. Trans Assoc Amer Phys 86:333–339
526. Wynne-Davies's Wynne-Davies's R (1973) Heritable disorders in orthopaedic practice. New Eng J Med 289:928, Review
527. Walker BA, McKusick VA (1973) The genetic basis of intrinsic diseases of the bone, in Progress of Pediatric Radiology. S. Karger A.G, Basel, Switzerland
528. McKusick VA (1973) The state of the department. In: Speech presented at annual meeting, Department of Medicine, J.H.U. School of Medicine, 25 Sept 1973
529. McKusick VA, Martin GR, Lichtenstein JR, Penttinen RPK, Rowe DW, Sussman MD, Byers PH (1973) Acquired and heritable defects in collagen synthesis and fibrogenesis. Trans Clin Climat Assoc 85:130–141
530. Borgaonkar DS, McKusick VA, Farber PA (1973) An inherited small extra chromosome: a mother with 46, XX, t (17;22) (pl;ql) and a son with 47, XY,+der(22) mat. J Med Genet 10:379–384
531. Melhelm R, Dorst JP, Scott CE, McKusick VA (1973) Roentgen findings in mucolipidosis III (pseudo-Hurler polydystrophy). Radiology 106:153–160
532. McKusick VA (1973) Phenotypic diversity of human diseases resulting from allelic series. Am J Hum Genet 25:446–456
533. McKusick VA (1973) Genetics and dermatology, or if I were to rewrite Cockayne's inherited abnormalities of the skin. J Invest Derm 60:343–359
534. Lichtenstein JR, Martin GR, Kohn LD, Byers PH, McKusick VA (1973) Defect in conversion of procollagen to collagen in a form of Ehlers-Danlos syndrome. Science 182:298–300
535. McKusick VA (1973) Ascertainment of genetic disease in relatives of affected persons. Israel J Med Sci 9:1358–1360
536. McKusick VA (1973) Summation: enzyme therapy in genetic disease. Birth Defects 9(2):229–321
537. McKusick VA (1973) Ethnic distribution of disease in non-Jews. Israel J Med Sci 9:1375–1382
538. McKusick VA (1973) Genetic aspects of epidemiology and preventive medicine. In: Sartwell PE (ed) Maxcy-Rosenau preventive medicine and public health, 10th edn. Appleton-Century-Crofts, New York, pp 591–601

539. Penttinen R, Lichtenstein J, Byers PH, Sussman MD, Rowe DW, McKusick VA, Martin GR (1974) Disorders identified as inherited defects in collagen: osteogenesis imperfecta. In: (Abstract) paper presented at the meeting of International Workshop on Calcified Tissues, at Gesher Haziv, Israel, 24–28 Mar 1974
540. McKusick VA (1974) Biomedical writing seminar I. Unpublished lecture, Hurd Hall, Johns Hopkins Hospital
541. McKusick VA (1974) Genetics and human disease, pp 323–341. Diseases of the aorta, pp 1247–1249. Inherited disorders of connective tissue, pp 2015–2023, in Harrison's Harrison's Principles of Internal Medicine, vol 7. New York, McGraw-Hill
542. McKusick VA (1974) Genetics and large bowel cancer. Am J Dig Dis 19:954–958
543. Wennstrom J, Pierce ER, McKusick VA (1974) Hereditary benign and malignant lesions of the large bowel. Cancer 34(Suppl):850–857
544. Jorgenson RJ, Levin LS, McKusick VA (1974) Heritable oral handicaps. Dent Clin N Am 18:579–594
545. Sussman MD, Lichtenstein JR, Nigra TP, Martin GR, McKusick VA (1974) Hydroxylysine-deficient skin collagen in a patient with a form of the Ehlers-Danlos syndrome. J Bone Joint Surg 56:1228–1234
546. McKusick VA, Frezal J (1974) Proceedings: report on the committee on terminology. Cytogenet Cell Genet 13:63–65
547. Kopits SE, Lindstrom JA, McKusick VA (1974) Pseudoachondroplastic dysplasia: pathodynamics and management. Birth Defects 10(12):341–352
548. Siggers CD, Rimoin DL, Dorst JP, Doty SB, Williams BR, Hollister DW, Silberberg R, Cranley RE, Kaufman RL, McKusick VA (1974) The Kniest syndrome. Birth Defects 10(9):193–208
549. Blizzard RM, McKusick VA (1974) Tolbutamide and achondroplasia. JAMA 228:1368, Letter
550. Hussels IE, Eikman EA, Kenyon KR, McKusick VA (1974) Treatment of mucopolysaccharidoses. Birth Defects 10(12):212–225
551. McKusick VA (1974) Works and aims of the Joseph Earle Moore Clinic. Johns Hopkins Med J 135:291
552. Penttinen RP, Lichtenstein JR, Martin GM, McKusick VA (1974) Abnormal collagen metabolism in cultured cells in osteogenesis imperfecta. Proc Nat Acad Sci 72:586–589
553. McKusick VA (1974) Multiple forms of the Ehlers-Danlos syndrome. Arch Surg 109:475–476, Editorial
554. McKusick VA (1974) Nomenclature of syndromes. Birth Defects 10(7):61–63
555. McKusick VA (1974) Discussion of classification. Birth Defects 10(9):193–208
556. Meyers KR, Golomb HM, Hansen JL, McKusick VA (1974) Familial neuromuscular disease with "myotubes". Clin Genet 5:327–337
557. McKusick VA (1974) Genetic counseling–hereditary conditions. In: Moghissi KS (ed) Birth defects and fetal development: endocrine and metabolic factors. Charles C Thomas, Springfield, Ill, pp 234–267
558. McKusick VA et al (1975) Mendelian inheritance in man catalogs of autosomal dominant, autosomal recessive and X-linked phenotypes. Johns Hopkins University Press, Baltimore
559. Borgaonkar DS, Bolling DR, Partridge C, Ruddle FH, McKusick VA (1975) Chromosomal variation in man: catalog of chromosomal variants and anomalies. Birth Defects 11(3):82–84
560. Rogers JG, McKusick VA (1975) Dominant familial arthritis with scoliosis. Birth Defects 11(6):75–80
561. McKusick VA (1975) Molecular defects in collagen. Ann Intern Med 82:585–586
562. Lichtenstein JR, McKusick VA (1975) Ehlers-Danlos syndrome with aortic aneurysm. Birth Defects 11(2):340
563. Kelly TE, Thomas GH, Taylor HA Jr, McKusick VA, Sly WS, Glasser JH, Robinow M, Luzzatti L, Espiritu C, Feingold M, Bull MJ, Asenhurst EM, Ives EJ (1975) Mucolipidosis III (pseudo-Hurler polydystrophy): clinical and laboratory studies in a series of 12 patients. Johns Hopkins Med J 137:156–175

564. Kelly TE, Thomas GH, Taylor HA, McKusick VA (1975) Mucolipidosis III: clinical and laboratory findings. Birth Defects 11(6):295–299
565. McKusick VA, Edwards JH (1975) Report of the committee on unassigned syntenic groups and theoretical considerations. Cytogenet Cell Genet 14:196–198
566. Kelly TE, Chase GA, Kaback MM, Kumor K, McKusick VA (1975) Tay-Sachs disease: high gene frequency in a non-Jewish population. Am J Hum Genet 27:287–291
567. Miller JD, McKusick VA, Malvaux P, Temtamy S, Salinas C (1975) The 3-M syndrome: a heritable low birthweight dwarfism. Birth Defects 11(5):39–47
568. McKusick VA (1975) The classification of heritable disorders of connective tissue. Birth Defects 11(6):1–9
569. Martin GR, Lichtenstein JR, McKusick VA, Penttinen R, Rowe DW, Pope FM, Sussman MD (1975) The molecular bases of certain inherited diseases of connective tissues involving collagen. Birth Defects 11(6):11–13
570. McKusick VA (1975) Presidential address: the growth and development of human genetics as a clinical discipline. Am J Hum Genet 27:261–273
571. McKusick VA (1975) Mendelian inheritane in man. Catalogs of autosomal dominant, autosomal recessive, and X-linked phenotypes, 4th edn. Johns Hopkins University Press, Baltimore
572. Siggers DC, Haciska DT, McKusick VA (1975) Vestibular dysfunction in familial dysautonomia. The Riley-Day syndrome. Arch Dis Child 50:890–893
573. McKusick VA (1975) A plan for reorganization of the Osler Medical Service. Johns Hopkins Med J 136:231–237
574. McKusick VA, Martin GR (1975) Molecular defects in collagen. Ann Int Med 82:585–586
575. Pope FM, Martin GR, Lichtenstein JR, Pettinen R, Gerson B, Rowe DW, McKusick VA (1975) Patients with Ehlers-Danlos syndrome type IV lack type III collagen. Proc Nat Acad Sci 72:1314–1316
576. McKusick VA, Ad Hoc Committee on Genetic Counseling (1975) Genetic counseling. Am J Hum Genet 27:240–242
577. McKusick VA (1975) The mapping of human chromosomes. In: Katz SH (ed) Biological anthropology. W.H. Freeman, San Francisco, pp 170–187
578. McKusick VA (1976) The Jeremiah Metzger lecture of the American clinical and climatological association: new genetic insight into old diseases. Trans Amer Clin Clim Assoc 87:193–213
579. Schleutermann CA, Murdoch JL, Walker BA, Bias WB, Chase GA, Friedhoff LB, McKusick VA (1976) A linkage study of the Marfan syndrome. Clin Genet 10:51–53
580. Hsu SH, Schacter BZ, Delaney DL, Miller TB, McKusick VA, Kenneth RH, Bodmer JG, Young D, Bodmer WF (1976) Genetic characteristics of the HeLa cell. Science 191:392–394
581. McKusick VA (1976) Max Broedel, in dictionary of scientific biography. Charles Scribner's Scribner's Sons, New York
582. McKusick VA (1976) Medical genetics. Hosp Pract II:27–28, Editorial
583. McKusick VA (1976) Osler as a medical geneticist. Johns Hopkins Med J 139:163–174
584. McKusick VA (1976) Pleiotropism. Am J Hum Genet 28:301–302, Letter
585. Stevenson RE, Howell RR, McKusick VA, Suskind JR, Hanson JW, Elliott DE, Neufeld EF (1976) The iduronidase-deficient mucopolysaccharidoses: clinical and roentgenographic features. Pediatrics 57:111–122
586. Murray HW, Mann JJ, Genecin A, McKusick VA (1976) Fever with dissecting aneurysm of the aorta. Am J Med 61:140–144
587. McKusick VA (1976) Medical genetics. In: Harvey AM, Johns RJ, Owens AH, Ross RS (eds) The principles and practice of medicine, 19th edn. Prentice-Hall, Appleton-Century-Crofts, New York, pp 533–601
588. McKusick VA (1976) The Lilly Lecture: heritable disorders of connective tissue: new clinical biochemical aspects. In: Peters DK (ed) Proceedings of the twelfth symposium on advanced medicine. Pitman Medical Publishing Co., Ltd, Kent, England, pp 170–191

589. McKusick VA, Klinger HP, Bootsma D, Ruddle FH (1976) Human gene mapping 3: Baltimore conference (1975). Third international workshop on human gene mapping. Birth Defects 12(7):1–452 (Also in Cytogenet. Cell Genet. 16: 1–452, 1976)
590. McKusick VA (1976) The aorta in the Marfan syndrome and other genetic disorders of connective tissue. In: Anagnostopoulos CE (ed) Lethal diseases of the ascending aorta. University Park Press, Baltimore, pp 113–123
591. Levin LS, Perin JCS, Ose L, Dorst JP, Miller JD, McKusick VA (1977) A heritable syndrome of craniosynostosis, short thin hair, dental abnormalities, and short limbs: cranioectodermal dysplasia. J Pediat 90:55–61
592. McKusick VA (1977) Foreword to Clinical Atlas of Human Chromosomes by J. DeGrouchy and C. Turleau. John Wiley, New York
593. Come PC, Bulkley BH, McKusick VA, Fortuin NJ (1977) Echocardiographic recognition of silent aortic root dilatation of Marfan's syndrome. Chest 72:789–792
594. Skovby F, McKusick VA (1977) Estrogen treatment of tall stature in girls with the Marfan syndrome. Birth Defects 13(3C):155–161
595. Gale AM, McKusick VA, Hutchins GM, Gott VL (1977) Familial congenital bicuspid aortic valve: secondary calcific aortic stenosis and aortic aneurysm. Chest 72:668–670
596. Pope FM, Martin GR, McKusick VA (1977) Inheritance of Ehlers-Danlos syndrome type IV syndrome. J Med Genet 14:200–204
597. Amrhein JA, Klingensmith GJ, Walsh PC, McKusick VA, Migeon CJ (1977) Partial androgen insensitivity: the Reifenstein syndrome revisited. New Eng J Med 297:350–356
598. Ose L, McKusick VA (1977) Prophylactic use of propranolol in the Marfan syndrome to prevent aortic dissection. Birth Defects 13(3C):163–169
599. McKusick VA, Ruddle FH (1977) The status of the gene map of the human chromosomes. Science 196:390–405
600. Gale AN, Lacassie Y, Rogers JG, Levin LX, McKusick VA (1977) Two "new" autosomal recessive mental retardation syndromes observed among the Amish. Birth Defects 13(3B):127–138
601. McKusick VA (1977) Genetic disorders of supporting tissue. In: Thorn GW, Adams RD, Braunwald E, Isselbacher KJ, Petersdorf RJ (eds) Harrison's Harrison's principles of internal medicine, 8th edn. McGraw-Hill, Inc., Blakiston Publication, New York, pp 2081–2088
602. Lacassie Y, Sack GH Jr, McKusick VA (1977) An autosomal dominant form of arthrogryposis multiplex congenita (AMC) with unusual dermatoglyphics. Birth Defects 13(3B):246–247, Abstract
603. Hamerton JL, Klinger HP, McKusick VA, Evans J (1978) Human gene mapping 4 (Winnipeg conference). Birth Defects 14(4):1–730 (Also in Cytogenet. Cell Genet. 22: 1–730, 1978)
604. McKusick VA (ed) (1978) Medical genetic studies of the amish: selected papers. Johns Hopkins University Press, Baltimore
605. McKusick VA (1978) Mendelian inheritance in man: catalogs of autosomal dominant, autosomal recessive and X-linked phenotypes, 5th edn. Johns Hopkins University Press, Baltimore
606. Greene HL, Ryan PF, McKusick VA (1978) Profound first-degree atrioventricular block: a 30-year study. Chest 74:212–214
607. McKusick VA (1978) The birth defects conferences: the last ten years and the next. Birth Defects 14(6B):161–170
608. Edwards JH, McKusick VA (1978) Report of the committee on unassigned linkage groups. (4th Human Gene Mapping Workshop). Cytogenet Cell Genet 22:129–131
609. Temtamy SA, McKusick VA (1978) The genetics of hand malformations, vol I, 14th edn, Birth Defects. Alan R. Liss, New York
610. McKusick VA, Neufeld E, Kelly TE (1978) The mucopolysaccharide storage diseases. In: Stanbury JB, Wyngaarden JB, Fredrickson DS (eds) The metabolic basis of inherited disease, 4th edn. McGraw-Hill, New York, pp 1282–1307
611. McKusick VA (1978) Genetic studies in inbred populations with particular reference to the Old Order Amish. In: Cohen BH, Lilienfeld AM, Huang PC (eds) Genetic issues in public health and medicine. Charles C Thomas, Springfield, Ill, pp 311–326

612. McKusick VA (1978) The William Allan memorial award lecture: genetic nosology: three approaches. Am J Hum Genet 30:105–122
613. McKusick VA (1979) No graying of grand rounds at Hopkins. New Eng J Med 300:440, Letter
614. Pyeritz R, McKusick VA (1979) The Marfan syndrome: diagnosis and management. New Eng J Med 300:772–777
615. Omenn GS, McKusick VA (1979) The association of Waardenburg syndrome and Hirschsprung megacolon. Am J Med Genet 3:217–233:217–223
616. Sulica VI, Cooper PH, Pope FM, Hambrick GW Jr, Gerson BM, McKusick VA (1979) Cutaneous histologic features in Ehlers-Danlos syndrome: study of 21 patients. Arch Dermatol 115:40–2115:40–42
617. Pyeritz RE, Murphy EA, McKusick VA (1979) Clinical variability in the Marfan syndrome. Birth Defects 15:155–178
618. McKusick VA, Edwards JH (1979) Report of the committee on unassigned linkage groups (5th Human Gene Mapping Workshop). Cytogenet Cell Genet 25:74–76
619. Evans HJ, Hamerton JL, Klinger HP, McKusick VA (1979) Human gene mapping 5 (Edinburgh conference). Birth Defects 15(II):1–236, Also, Cytogenet. Cell Genet. 25: 1–236, 1979
620. Harvey AM, Johns RJ, McKusick VA, Owens AH Jr, Ross RS (eds) (1980) The principles and practice of medicine, 20th edn. Appleton-Century-Crofts, New York
621. McKusick VA (1980) The genetics of birth defects. Ann Rev Birth Defects : 229
622. McKusick VA (1980) The anatomy of the human genome. Am J Med 60:267–276
623. McKusick VA (1980) The Amish. Endeavour 4:52–57
624. O'Donnell O'Donnell FE, Green WR, McKusick VA, Forsius H, Eriksson AW (1980) Forsius-Eriksson syndrome: its relation to the Nettleship-Falls X-linked ocular albinism. Clin Genet 17:403–408
625. McKusick VA (1980) Birth Defects 16:12
626. McKusick VA (1980) A short history of medical genetics. In:Kelly TE (ed) Surgical forum, Yearbook Medical Publishers, Inc., pp 1–22
627. Steinmann B, Tuderman L, Peltonen L, Martin GR, McKusick VA, Prockop DJ (1980) Evidence for a structural mutation of procollagen type I in a patient with the Ehlers-Danlos syndrome type VII. J Biol Chem 18:8887–8893
628. McKusick VA (1980) The anatomy of the human genome: the Wilhelmine E. Key 1979 invitational lecture. J Hered 71:370–391
629. McKusick VA, Edwards JH (1980) Report of the committee on unassigned linkage groups. Birth Defects 15:74–76
630. McKusick VA (1980) The human genome through the eyes of Mercator and Vesalius. Trans Am Clin Climatol Assn 92:66–90
631. McDonald GR, Schaff HV, Pyeritz RE, McKusick VA, Gott VL (1981) Surgical management of patients with the Marfan syndrome and dialation of the ascending aorta. J Thorac Cardiovasc Surg 81:180–186
632. Kwiterovich PO Jr, Smith HH, Connor WE, Bachorik PS, McKusick VA, Teng B, Sniderman AD (1981) Hyperapobetalipoproteinaemia in two families with xanthomas and phytosterolaemia. Lancet 1:466–469
633. Trojak JE, Polmar SH, Winkelstein JA, Hsu S, Francomano C, Pierce GF, Scillian JJ, Gale AN, McKusick VA (1981) Immunologic studies of cartliage-hair hypoplasia in the Amish. Johns Hopkins Med J 148:157–164
634. McKusick VA (1981) The genetics of birth defects: the gene map of the human chromosomes in relation to diagnosis and management. Birth Defects 17:229–248
635. McKusick, V.A.: Contributions of clinical genetics to human genetics, in Problems in General Genetics: Proceedings of the XIV Intl. Congress of Genetics, Vol. II, Book I. Moscow: MIR. Pub, 1981, pp. 170–176.
636. McKusick VA (1981) The last twenty years: an overview of advances in medical genetics. In: Mammalian genetics and cancer: the Jackson laboratory fiftieth anniversary symposium. Alan R. Liss, Inc, New York, pp 127–144

637. McKusick VA (1981) Principles and applications of medical genetics. Hi-Medic 2:2–7
638. McKusick VA (1981) The anatomy of the human genome. Hosp Pract 16(4):82–100
639. Cupo LN, Pyeritz RE, Olson JL, McPhee SJ, Hutchins GM, McKusick VA (1981) Ehlers-Danlos syndrome with abnormal collagen fibrils, sinus of valsalva aneurysms, myocardial infarction, panacinar emphysema and cerebral heterotopias. Am J Med 71:1051–1058
640. Lehninger AL, McKusick VA, Santora PB (1981) Proceedings of the conference on genetic and biochemical variability in response to alcohol. Alcohol Clin Exp Res 5:436–460
641. Mudd HH, Havlik R, Levy HL, McKusick VA, Feinleib M (1981) A study of cardiovascular risk in heterozygotes for homocystinuria. Am J Hum Genet 33:883–893
642. Pyeritz RE, McKusick VA (1981) Basic defects in the Marfan syndrome. New Eng J Med 305:1011–1012, Editorial
643. McKusick VA (1982) Genetic disorders of the human nervous system: a commentary. In: Schmitt FO (ed) Molecular genetic neuroscience. Raven Press, New York
644. McKusick VA (1982) Clinical genetics: a neo-Vesalian view of the past and the future. In: Bonne-Tamir B (ed) Human genetics, part B: medical aspects. Alan R. Liss, New York
645. McKusick VA (1982) Genetics and cardiovascular disease: 1981. Jap Circ J 46:1–7
646. McKusick VA (1982) The human genome through the eyes of a clinical geneticist. Cytogenet Cell Genet 32:7–23
647. Mudd SH, Havlik R, Levy HL, McKusick VA, Feinleib M (1982) Cardiovascular risk in heterozygotes for homocystinuria. Am J Hum Genet 34:1018–1021, Letter
648. McKusick VA (1983) Foreword. In: Connor JM (ed) Soft tissue ossification. Springer-Verlag, Berlin
649. McKusick VA (1983) Mendelian inheritance in man. Catalogs of autosomal dominant, autosomal recessive and X-linked phenotypes, 6th edn. Johns Hopkins University Press, Baltimore
650. Come PC, Fortuin NJ, White RI Jr, McKusick VA (1983) Echocardiographic assessment of cardiovascular abnormalities in the Marfan syndrome. Am J Med 74:465–474
651. Kelly TE, McKusick VA (1984) Mucopolysaccharidosis I-H, in current ocular therapy, 2nd edn. W.B. Saunders Co, Philadelphia
652. McKusick VA, Conneally PM (1984) Report of the Committee on Unassigned Linkage Groups (7th Gene Mapping Workshop). Cytogenet Cell Genet 37:205–209
653. McKusick VA (1984) The human gene map 15 November 1983. Clin Genet 25:89–123
654. Renie WA, Murphy RP, Anderson KC, Lippman SM, McKusick VA, Proctor LR, Shimizu H, Patz A, Fine SL (1984) The evaluation of patients with Eales' disease. Retina 13:1276–1277
655. McKusick VA (1984) The William Allan Memorial Award: Introduction of Dr. Frank Ruddle. Am J Hum Genet 36:941–953
656. McKusick VA (1984) Sarcoidosis – a case study in nosology. In: Address before the sixth international conference on sarcoidosis. Baltimore
657. McKusick VA (1984) Diseases of the genome. JAMA 252:1041–1048, Interview
658. Cohen BH, McKusick VA (1984) Abraham M. Lilienfeld: in memoriam. Genetic Epidemiol 1:297–300
659. Harvey AM, Johns RJ, McKusick VA, Owens AH Jr, Ross RS (eds) (1984) Principles and practice of medicine, 21st edn. Appleton-Century-Crofts, New York
660. McKusick VA (1985) The human gene map 1 December 1984. Clin Genet 27:207–239
661. McKusick VA (1985) Marcella O'Grady Boveri (1865–1950) and the chromosome theory of cancer. J Med Genet 22:431–440
662. McKusick VA, Pyeritz RE (1985) Heritable and developmental disorders of connective tissues and bone. In: McCarty DJ (ed) Arthritis and allied conditions, 10th edn. Lea & Febiger, Philadelphia, pp 1168–1194
663. McKusick VA (1986) The human gene map 15 April 1986. Clin Genet 29:545–548
664. McKusick VA (1986) Foreword. In: Ricardi V, Eichner JE (eds) Neurofibromatosis: phenotype, pathogenesis, and natural history. Johns Hopkins University Press, Baltimore
665. McKusick VA (1986) The Gene Map of Homo sapiens: Status and Prospectus. Cold Spring Harbor Symp. Quant Biol 51:15–27, Appendix

666. McKusick VA (1986) Mendelian inheritance in man. Catalogs of autosomal dominant, autosomal recessive, and X-linked phenotypes, 7th edn. Johns Hopkins University Press, Baltimore
667. McKusick VA (1987) Correction: telangiectases, not telangiectasias. Letter to the editor. Ann Int Med 106:336
668. McKusick VA, Roderick TH, Mori J, Paul NW (eds) (1987) Medical and experimental mammalian genetics: a perspective. Alan R. Liss, Inc., New York
669. Offerhaus GJA, Levin LS, Giardiello FM, Krush AJ, Welsh SB, Booker SV, Hasler JF, McKusick VA, Yardley JH, Hamilton SR, Luck GD (1987) Occult radiopaque jaw lesions in familial adenomatous polyposis coli and hereditary nonpolyposis colorectal cancer. Gastroenterology 93:490–497
670. McKusick VA, Ruddle FH (1987) Toward a complete map of the human genome. Genomics 1:103–106, Editorial
671. McKusick VA, Ruddle FH (1987) A new discipline, a new name, a new journal. Genomics 1:1–2
672. McKusick VA (1987) Molecular biology of homo sapiens. Genomics 1:196–200, Review
673. McKusick VA, Maroteaux P (1987) Committee on clinical disorders. Cell Genet 46:325–38, 46:325–338, (Ninth Human Gene Mapping Workshop) Cytogenet
674. Hall JG, Dorst JP, Rotta J, McKusick VA (1987) Gonadal masaicism in pseudoachondroplasia. Am J Med Genet 28:143–5128:143–151
675. McKusick VA, Roderick TH (1987) Twenty-five years in medical genetics and experimental mammalian genetics with particular reference to the gene map of mouse and man. Birth Defects 23:1–17
676. McKusick VA (1988) The new genetics and clinical medicine: a summing up. Hosp Pract 23:177–83, 23:177–183, 186, 191
677. McKusick VA (1986) The morbid anatomy of the human genome: a review of gene mapping in clinical medicine (four parts). Medicine 65:1–33, 66:1–63, 1987; 66: 237–296, 1987; 67:1–19, 1988
678. National Research Council (1988) Committee on mapping and sequencing the human genome: mapping and sequencing the human genome. National Academy Press, Washington, D.C
679. Harvey AM, Johns RJ, McKusick VA, Owens AH Jr, Ross RS (eds) (1988) Principles and practice of medicine, 22nd edn. Appleton and Lange, Norwalk Ct
680. McKusick VA (1988) Mendelian inheritance in man. Catalogs of autosomal dominant, autosomal recessive and X-linked phenotypes, 8th edn. The Johns Hokins University Press, Baltimore
681. Horton WA, Goetnick PF, Sayedin SM, Minor RR, Spranger J, Rimoin DL, McKusick VA, Dayton DH (1988) New perspectives in the human chondrodysplasias. Patho Immunopathol Res 7:146–87:146–148
682. Beighton P, de Paepe A, Danks D, Finidori G, Gedde T, Goodman R, Hall JG, Hollister DW, Horton W, McKusick VA et al (1988) International nosology of heritable disorders of connective tissue, Berlin, 1986. Am J Med Genet 29:581–9429:581–594
683. McKusick VA (1989) Forty years of medical genetics. JAMA 261:3155–3158
684. McKusick VA (1989) Mapping and sequencing the human genome. New Eng J Med 320:910–915
685. Harvey AM, Brieger GH, Abrams SL, McKusick VA (1989) A model of its kind. A centennial history of medicine at johns hopkins, vol I. Johns Hopkins Press, Baltimore
686. Harvey AM, Brieger GH, Abrams SL, Fishbein JM, McKusick VA (1989) A model of Its Kind, vol 2, A pictorial history of medicine at Johns Hopkins. Johns Hopkins University Press, Baltimore
687. McKusick VA (1989) Historical perspectives: the understanding and management of genetic disorders. Md Med J 38:901–908
688. McKusick VA (1989) HUGO news: the human genome organization: history, purposes and membership. Genomics 5:385–387

689. McKusick VA (1990) Mapping and sequencing the human genome: status, strategies, prospects, and implications. Jpn J Human Genet 35:15–16
690. McKusick VA (1990) The morbid anatomy of the human genome: the role of gene mapping in clinical medicine. Acta Paediatr Jpn 34:234–241
691. McKusick VA (1990) Mendelian inheritance in man. Catalogs of autosomal dominant, autosomal recessive and X-linked phenotypes, 9th edn. Johns Hopkins University Press, Baltimore
692. McKusick VA (1991) Abraham Lincoln and Marfan syndrome. Nature 352:280
693. McKusick VA (1991) The defect in Marfan syndrome. Nature 352:279–281, News and Views
694. Gott CL, Pyeritz EW, Cameron DE, Greene PS, McKusick VA (1991) Composite graft repair of Marfan aneurysm of the ascending aorta: results in 100 patients. Ann Thorac Surg 52:38–45
695. McKusick VA (1991) Genomic mapping and how it has progressed. Hosp Pract 26:74–84
696. McKusick VA (1991) The human genome project and clinical medicine. Hosp Pract 26:15–16, Editorial
697. Levitt RC, Nouri N, Jedlicka AE, McKusick VA, Marks AR, Shutack JF, Fletcher JE, Rosenberg H, Meyers DA (1991) Evidence of genetic heterogenetiy in malignant hyperthermia susceptibility. Genomics 11:543–547
698. Schwindinger WF, Francomano CA, Levin MA, McKusick VA (1991) DNA light on the Tegernsee giant (letter). Lancet 338:1454–5338:1454–1455
699. McKusick VA (1992) Morbid anatomy of the human genome in relation to hematological disorders: Role of gene mapping in hematology. In: Bartsocas CS, Loukopoulos D (eds) Genetics of hematological disorders. Hemisphere Publishing Corp, New York, pp 3–23
700. McKusick VA (1992) Human genetics: the last 35 years, the present, and the future. Am J Hum Genet 50:663–7050:663–670
701. McKusick VA (1992) DNA technology in forensic science. Committee on DNA technology in forensic science. Board on biology; commission on life science; national research council. National Academic Press, Washington, DC
702. McKusick VA (1992) Mendelian inheritance in man. Catalogs of autosomal dominant, autosomal recessive and X-linked phenotypes, vol 2, 10th edn. The Johns Hopkins University Press, Baltimore
703. McKusick VA, McKusick VA (1992) The human genome project: plans, status, and applications in biology and medicine. In: Annas GJ, Elias S (eds) Gene mapping. Using law and ethics as guides. Oxford University Press, New York, pp 18–42
704. Beighton P (ed) (1992) McKusick's McKusick's heritable disorders of connective tissue, 5th edn. Mosby, St. Louis
705. McKusick VA (1993) Heritable disorders of connective tissue: a personal account of the origins, evolution, validation, and expansion of a concept. In: Royce PM, Steinmann B (eds) Connective tissue and its heritable disorders. Wiley-Liss, New York
706. McKusick VA (1993) Medical genetics: a 40-year perspective on the evolution of a medical specialty from a basic science. JAMA 270:2351–2356
707. McKusick VA (1993) Medical genetics: a self-instruction guide and workbook based on mendelian inheritance in man. Johns Hopkins University Press, Baltimore
708. McKusick VA (1994) Alkaptonuria: tracked down to chromosome 3. Genomics 19:3–4
709. Kucherlapati RS, McKusick VA, Ruddle FH (1994) Editorial. Genomics 19:1–2
710. Lebwohl M, Neldner K, Pope FM, De Paepe A, Christiano AM, Boyd CD, Uitto J, McKusick VA (1994) Classification of pseudoxanthoma elasticum: report of a consensus conference. J Am Acad Dermatol 30:103–107
711. McKusick VA, Amerger JS (1994) The morbid anatomy of the human genome: chromosomal location of mutations causing disease (update 1 Dec. 1993). J Med Genet 3:265–279
712. Tugendreich S, Bassett DE Jr, McKusick VA, Boguski MS, Hieter P (1994) Genes conserved in yeast and humans. Hum Molec Genet 3:1509–1517
713. Romeo G, McKusick VA (1994) Phenotypic diversity, allelic series and modifier genes. Nature Genet 7:451–453

714. Gott VL, Cameron DE, Pyeritz RE, Gillinov AM, Greene PS, Stone CD, Alejo DE, McKusick VA (1994) Composite graft repair of Marfan aneurysm of the ascending aorta: result in 150 patients. J Card Surg 9:482–489
715. Gott VL, Gillinov AM, Pyeritz RE, Cameron DE, Reitz BA, Greene PS, Stone CD, Ferris RL, Alejo DE, McKusick VA (1994) Aortic root replacement: risk factor analysis of a 17 year experience with 270 patients. Presented at the 74th annual meeting of the American Association for Thoracic Surgery, , New York, 27 Apr 1994
716. Sulisalo T, Francomano CA, Sistonen P, Maher JF, McKusick VA, DeLaChapelle A, Kaitila I (1994) High-resolution genetic mapping of the cartilage-hair hypoplasia (CHH) gene in Amish and Finnish families. Genomics 20:347–353
717. Rosenfeld PJ, McKusick VA, Amberger JS, Dryja TP (1994) Recent advances in the gene map of inherited eye disorders: primary hereditary diseases of the retina, choroid, and vitreous. J Med Genet 31:903–915
718. Antonarakis SE, McKusick VA (1994) Discussion on mutation nomenclature (Letter). Hum Mutat 4:166
719. Wilkie AO, Amberger JS, McKusick VA (1994) A gene map of congenital malformations. J Med Genet 31:507–517
720. McKusick VA (1994) Mendelian inheritance in man. Catalogs of human genes and genetic disorders, vol 2, 11th edn. Johns Hopkins University Press, Baltimore
721. McKusick VA (1995) Reviews in molecular medicine (Editorial). Medicine 74:301–304
722. McKusick VA et al. OMIM (online Mendelian Inheritance in Man). URL: http:/www.ncbi.nlm.nih.gov/omim/
723. McKusick VA (1996) History of medical genetics. In: Rimoin DL, Connor JM, Pyeritz RE (eds) Emery-Rimoin principles and practice of medical genetics, 3rd edn. Churchill Livingston, Edinburgh, pp 1–30
724. McKusick VA, Amberger JS, Francomano CA (1996) Progress in medical genetics: map-based gene discovery and the molecular pathology of skeletal dysplasias. Am J Med Genet 63:98–105
725. Beutler R, Mckusick VA, Motulsky AG, Scriver CR, Hutchinson F (1996) Mutation nomenclature: nicknames, systematic names, and unique identifiers. Hum Mutat 8:203–206
726. Gott VL, Laschinger JC, Cameron DE, Dietz HC, Greene PS, Gillinov AM, Pyeritz RE, Alejo DE, Fleischer KJ, Anhalt GJ, Stone CD, McKusick VA (1996) The Marfan syndrome and the cardiovascular surgeon. Eur J Cardiothorac Surg 10:149–158
727. Kucherlapati RS, McKusick VA, Ruddle FH (1996) Genomics: an established discipline, a commonly used name, a mature journal (Editorial. Genomics 31:1–2
728. Rushton AR (1994) Genetics and medicine in the United States, 1800–1922. JHU Press, Baltimore, Bull. Hist. Med. 70:731–733, 1996
729. Goebel HH, Seddigh S, Hopf HC, Uemichi T, Benson MD, McKusick VA (1997) A European family with histidine 58 transthyretin mutation in familial amyloid polyneuropathy. Neuromuscul Disord 7:229–230
730. McKusick VA (1997) Molecular genetics in clinical practice: the series begins. Hosp Pract 32:11–12
731. Arch EM, Goodman BK, Van Wesep RA, Liaw D, Clarke K, Parsons R, McKusick VA, Geraghty MT (1997) Deletion of PTEN in a patient with Bannayan-Riley-Ruvalcaba syndrome suggest allelism with Cowden disease. Am J Med Gene 71:489–493
732. Howard TD, Guttmacher AE, McKinnon W, Sharma M, McKusick VA, Jabs EW (1997) Autosomal dominant postaxial polydactyly, nail dystrophy, and dental abnormalities map to chromosome 4p16, in the region containing the Ellis-van Creveld syndrome locus. Am J Hum Genet 61:1405–1412
733. McKusick VA (1997) Observation over 50 years concerning intestinal polyposis, Marfan syndrome and achondroplasia, 1997 Albert Lasker award for special achievement in medical science. Nature Medicine (Commentary) 3:1065–1068
734. McKusick VA (1997) Perspective. Genomics: structural and functional studies of genomes. Genomics 45:244–249

735. McKusick VA (1997) Postscript: organelle diseases. In: Applegarth DA, Dimmick JE, Hall JG (eds) Chapman and Hall Medical, London
736. McKusick VA (1998) Mendelian inheritance in man. A catalog of human genes and genetic disorders. disorders, vol 3, 12th edn. The Johns Hopkins University Press, Baltimore (12th edn. In 3 vol)
737. Dietz HC, McKusick VA (1999) Inherited vascular disease. In: Chien RK (ed) Molecular basis of cardiovascular disease. A companion to Braunwald's Braunwald's heart disease. W.B. Saunders, Estados Unidos, pp 167–187
738. Smith FJD, McKusick VA, Nielsen K, Pfendner E, Uitto J, McLean WHI (1999) Cloning of multiple keratin 16 genes facilitates prenatal diagnosis of pachyonychia congenita Type I. Prenat Diagn 19:941–946
739. McKusick VA (1999) In memoriam. Abner McGehee harvey. 1911–1998. Proc Assoc Am Physicians 111(4):365–8111(4):365–368
740. McKusick VA, Naggert J, Nishina P, Valle D (1999) 40 years of the annual bar harbor course (1960–1999): a pictorial history. Clin Genet 55:398–415
741. McKusick VA (1999) A grade 6 systolic murmur. N Eng J Med 341(19):1472–73341(19): 1472–1473
742. Friedman JM, Gutmann DH, MacCollin M, Riccardi VM (eds) (1999) Neurofibromatosis: phenotype, natural history, and pathogenesis, vol 3. Johns Hopkins University Press, Baltimore
743. (1999) Review: Two lives of Osler, 75 years apart. The Baltimore Sun Paper
744. Srsen S, Srsenova K (2000) Foreword: Základy Klinickej Genetiky: A Jej Molekulárna Podstata. Vydavatelstvo Osveta, spol. s.r.o., Oslobditelov 21, 036 06 Martin, SR
745. McKusick VA (2000) Biographical memoirs. A. McGehee harvey, 30 July 1911–8 May, 1998. Proc Am Phil Soc 144:85–94
746. McKusick V (2001) Protector genes. In: Boulyjenkov V, Berg K, Christen Y (eds) Genes and resistance to disease. American Scientist, pp 85–86
747. Collins FS, McKusick VA (2001) Implications of the human genome project for medical science. JAMA 285:540–544
748. Peltonen L, McKusick VA (2001) Dissecting human disease in the postgenomic era. Science 291:1224–1229
749. Venter JC, Adams MD, Myers EW, Li PW, Mural RJ, Sutton GG, Smith HO, Yandell M, Evans CA, Holt RW, Gocayne JD, Amanatides P, Ballew RM, Huson DH, Wortman JR, Zhang Q, Kodira CD, Zheng XH, Chen L, Skupski M, Subramanian G, Thomas PD, Zhang J, Miklos GLG, Nelson C, Broder S, Clark AG, Nadeau J, McKusick VA et al (2001) The sequence of the human genome. Science 291:1304–1351
750. McKusick VA (2001) Mapping the human genome: retrospective, perspective, and prospective. In: Mahowald MB, McKusick VA, Scheuerle AS, Aspinwal TJ (eds) Genetics in the clinic: clinical, ethical, and social implications for primary care. Mosby, St. Louis
751. Mahowald MB, McKusick VA, Scheuerle AS, Aspinwall TJ (eds) (2001) Genetics in the clinic: clinical, ethical, and social implications for primary care. Mosby, St. Louis
752. McKusick VA (2001) Persisting memories of Cyril Clarke in Baltimore. J Med Gene 38:284
753. Watson JD (2002) A passion for DNA: genes, genomes and society. Bull Hist Med (in press)
754. McKusick VA (2001) The anatomy of the human genome. A Neo-Vesalian basis for medicine in the 21st century. JAMA 286:2289–2295
755. Hamosh A, Scott AF, Amberger J, Bocchini C, Valle D, McKusick VA (2001) Online mendelian inheritance in man (OMIM), a knowledgebase of human genes and genetic disorders. Nucleic Acids Res 30:1–4
756. McKusick VA (2002) Heritable disorders of connective tissue: a personal account of the origins, evolution, validation, and expansion of a concept. In: Royce PM, Steinmann B (eds) Connective tissue and its heritable disorders. Molecular, genetic, and medical aspects, 2nd edn. Wiley-Liss, New York

757. Mckusick VA (2002) History of medical genetics. In: Rimoin DL, Connor JM, Pyeritz RE, Korf BR (eds) Emery-rimoin principles and practice of medical genetics, 4th edn. Churchill Livingstone, Edinburgh, pp 3–36
758. Ridanpää M, Jain P, McKusick VA, Francomano CA, Kaitila I (2003) The major mutation in the RMRP gene causing CHH among the amish is the same as that found in most finnish cases. AJMG 121C:81–83
759. Francomano C, McKusick VA, Biesecker LC (2003) Medical genetic studies in the amish: historical perspective. AJMG 121C:1–4
760. McKusick VA (2004) The Marfan syndrome: from clinical delineation to mutational characterization, a semiautobiographic account. In: Robinson PN, Godfrey M (eds) Marfan syndrome: a primer for cllinicians and scientist. Kluwer Academic/Plenum, New York
761. McKusick VA (2004) From Karl Landsteiner to Peter Agre: 100 years in the history of blood group genetics. Transfusion 44:1370–1376
762. McKusick VA (2005) The Gordon Wilson lecture: the clinical legacy of Johnathan Hutchinson (1828 1913): syndromology and dysmorphology meet genomics. Trans Am Clin Climatol Assoc 116:15–38
763. McKusick VA (2006) A 60-year tale of spots, maps, and genes. Annu Rev Genomics Hum Genet 7:1–27
764. McKusick VA (2007) Mendelian inheritance in man and its online version OMIM. AJHG 80:588–604
765. McKusick VA (2007) 2006 Curt Stern award address. Introductory speech for Hal Dietz. AJHG 81:660–661
766. McKusick VA (2007) The clinical legacy of Jonathan Hutchinsom (1828–1913). Fifty years of human genetics. In: Mayo O, Leach C (eds) A festschrift and liber amicorum to celebrate the life and work of George Robert Fraser. Wakefield Press, S. ASA
767. McKusick VA Obituary. Johns Hilton Edwards. Nat Genet 39:1417
768. McKusick VA (2007) History of medical genetics. In: Rimoin DL, Connor JM, Pyeritz RE, Korf BR (eds) Emery and Rimoin principles and practice of medical genetics, 5th edn. Churchill Livingston, Edinburgh, pp 3–32
769. McKusick VA (2007) Foreword: principles and practice of medical genetics, 5th edn. Churchill Livingston, Edinburgh
770. McKusic VA, Amberger J (2007) The morbid anatomy of the human genome. A chromosome-by chromosome synopsis of disease-related genes an genetic loci. Emery and Rimoin principles and practice of medical genetics, 5th edn. Churchill Livingston, Edinburgh, pp 191–193
771. McCabe LL, McCabe ERB (2008) DNA: promise and peril. University of California Press, Berkeley
772. Cathey SS, Kudo M, Tiede S, Raas-Rothchild A, Braulke T, Beck M, Taylor HA, Canfield WM, Leroy JG, Neufeld EF, McKusick VA (2008) Molecular order in mucolipidosis II and III nomenclature. Am J Med Genet A 146A:512–513

Appendix B

Appendix B 1. Eulogy by Ken McKusick

Ken McKusick
1404 Berwick Road
Towson, MD
21204

On behalf of the family, I'd like to thank all of you for coming today to celebrate the life of my Dad.

I spent a great deal of time talking with Dad over the past couple of months. It was highly satisfying quality time and helped me come to terms with what was happening. I was going to make a list of all the important occurrences and milestones that he felt lucky about. However, that list became very long and I realized that it would take too much time. So I'm going to say just a few things that are most important to me.

First and foremost, Dad's love for my mother was always very apparent. They had their first date two weeks after she arrived at Hopkins to begin medical school, while he was an intern, in 1946. She's still here 62 years later. He'd describe her as his best friend, companion, critic and editor, and staunch supporter through all these years. They had very complementary qualities and interests. Mom was the one to go to when you needed help with math homework, and Dad could help more with history and biology. Both of them would continue to correct improper grammar well beyond childhood.

Dad's devotion to his work, to which you've heard several testimonies today, is perhaps even more apparent in terms of what he did *not* know.

He couldn't tell you who Barry Bonds is.

He couldn't name any actor who ever played James Bond.

He couldn't name any two NFL teams that ever played each other in the Super Bowl.

K.R. Dronamraju and C.A. Francomano (eds.), *Victor McKusick and the History of Medical Genetics*, DOI 10.1007/978-1-4614-1677-7,

He couldn't name any actor or character that played on *any* of the following TV shows: The Honeymooners, Star Trek, M*A*S*H, Cheers, or Seinfeld.

And...he couldn't name any member of the Beatles.

The fact that his work was Dad's greatest passion was apparent in all aspects of his life. His hobbies were his work. After personally accumulating closets full of games and sports equipment, I realize that my dad never bought himself a toy, not a pool table, fishing equipment, golf clubs, sports car, boat, nor electronic gadget. Only when my parents moved into Blakehurst did they pay for cable TV (or get air conditioning). The closest I could come to toys were his camera and computer, both of which were essential for his work.

Nonetheless, Dad did a lot of things just because he loved us. He took my brother and me to the last game of the 1971 World Series despite the fact that he would not have remembered a thing about that game a year later. He drove us to little league, was active in scouting, and would participate in the family games we'd have most summer nights in Nova Scotia, although he always had a pile of reading close at hand. And then, there was the memorable overnight canoe trip Dad and I took down the Annapolis River in Nova Scotia.

Dad told me he was very proud of the fact that he never paid for an overseas trip after his honeymoon. Upon cross-examination that proved to be an exaggeration, he certainly did give us kids the opportunity to travel and experience the world with him. Many of our fondest memories came from these trips. In particular, my brother and I will never forget being pulled out of kindergarten and first grade, respectively, in 1970, to travel around the world via Fiji, Australia, New Zealand, Hong Kong, India, Iran, and England for two full months. On other trips, he took us along to conferences or speaking engagements in the Netherlands, Austria, Yugoslavia, and the Soviet Union at the height of the cold war.

I never thought we'd have a more special family trip, until this year when Mom and Dad took his twin brother, Vincent, Aunt Nancy, Victor, Lori, Maureen, and me to Japan. This trip afforded him a wonderful opportunity to meet with old colleagues, spend time with family, celebrate his life's work, and provided something to look forward to as his health declined. What made the trip so special for me wasn't just seeing him win the award but seeing how happy he was to share the experience with family and friends.

Most importantly, I'm grateful to both Dad and Mom, for their unflinching willingness to provide educational opportunities. None of us...Carol, Victor, or I... found our path immediately, but Mom and Dad were always willing to support us on our next goal.

As Dad spoke more about his life, I got a picture of a man thoroughly comfortable with who he was and how he lived. He was not without regrets (as he'd say, "People without regrets just aren't facing them"), but he was honest with himself, brave, mentally sharp, and cheerful to the end. No one...could ask for anything more.

Appendix B 2. Eulogy by Victor McKusick

Remembering My Dad

The Rev. Victor W. McKusick
933 Westwood Dr.
Herkimer, NY 13350

Pingree Cemetery
Parkman, Maine
August 8, 2008
11:00 AM

I don't know when I've been more proud to be my father's son.

Over the last month, I've been thinking about life and death and about my dad as he straddled that very tenuous line. Also during that time, I rented the recent film: "The Bucket List." The Bucket List is a tale of two men ailing with cancer in the twilight stages of their disease. The two men head off on an adventure of doing the things they have always wanted to do in life. I think of this film not in the sense that dad should have had great regrets about his life. To the contrary, Victor A. McKusick was a man who lived a full and rich life.

Dad never went on a vacation or a tourist junket. Even trips to Nova Scotia were just a wonderful change of venue to work or to exercise a different kind of work. All the "once in a lifetime" things he did and all the extraordinary places he saw were connected to his lifetime work in genetics and medicine. He never went skydiving as the Jack Nicholson and Morgan Freeman characters in the movie. However, I do remember seeing pictures of him in his bathing suit floating in the Dead Sea. This, of course, took place on a visit to Israel, while visiting with his colleague Dr. Goodman and meeting with the Israeli genetics community.

- He walked on the Great Wall of China on three separate occasions.
- Watched ski jumping competitions in Norway.
- As Ken noted last week, visited the Soviet Union at the height of the Cold War.
- Saw the Lipizzaner Stallions in Vienna
- Heard the Vienna Boys' Choir not in concert but street children in the opera Carmen.
- Not only visited Tito's Yugoslavia but had good family friends whom my brother and I referred to as Aunt Lilliana and Uncle Kreshau.
- And of course dined with the Emperor and Empress of Japan, also on a medical occasion, when he received the prestigious Japan Prize.

These would be "Bucket List experiences" for most of us; for dad, they were happenstance activities on his lifetime journey.

I think an even more poignant part of this motion picture, when I think of dad's life. In the scene, the two crusty characters are perched high above the Valley of the

Kings atop one of the pyramids. The character Carter says to Jack Nicholson character, "The ancient Egyptians believed one was asked two questions at the gates of heaven: Did you experience joy in life? The second question was did you bring joy to others?" And unlike the wealthy hedonistic Jack Nicholson character, who responds, "I guess I'm not getting into Egyptian Heaven then," by those standards, Victor A. McKusick is unequivocally, absolutely getting into Egyptian heaven. It is this joy and delivery of joy that I find in my memories of my dad.

I have to say I don't remember hearing many stories from dad about his childhood except the occasional twin story such as the switching of ties with Uncle Vincent at school to confuse their teachers. However, one childhood story was one of dad's favorite jokes. Grandma McKusick was busy putting the twins to bed one evening. She was having a particularly hard time in settling them down. "Come on know boys its time to go bed. I have other fish to fry." Supposedly, my father responded, "I want to stay up and have the fish." Such was Dad's sense of humor.

I'm finding it very difficult to encapsulate mom and dad's courtship and their 59-year marriage. Mom and dad went on their first date sometime in the late 1940s while they were both at Hopkins. They had dinner together and evidently ordered vanilla ice cream for dessert. My dad stated boldly, as to impress this very attractive woman: "I'll have ketchup on my ice cream. That's the way I like it."

I was reading an interview that described dad's career before we went to Japan. In the interview, both dad and mom spoke of going through Sir Dr. Jonathan Hutchinson's (personal physician to Queen Victoria) drawings in the basement of the Welch Library at Hopkins. The interviewer noted, "How romantic." For them, I'm sure it was. Although different specialties, mom was a partner in dad's practice of medicine.

There were two very loving acts that my father performed occasionally throughout my parent's marriage. They seemed to be small and insignificant yet were filled with the love he shared with my mother. I have tried to duplicate them in my own marriage. My dad developed the habit of bringing tea to mom in bed each morning. Although I don't bring tea to my wife every morning in bed, we have developed the ritual of my always asking as we prepare breakfast together if Lori wants orange juice. Furthermore, dad would frequently wash dishes with mom or make Ken and I do them. In the McKusick household in Herkimer, Lori and I do the dishes together after supper whenever we can.

Dr. Andrea Superti-Furga is a young physician my parents first met in Sestri Levante in 1988. I thought he captured dad's relationship to mom best when he wrote, "VAM was also a role model for the way he took Mrs. Anne with him as an eye level partner and affectionate spouse, not as the attachment-wife of a famous man."

Although she is not here, I wanted to share words from my sister Carol in memory of my dad. One time, Carol and I were looking at the black and white framed photo of dad sitting in a chair reading a book. Carol remembered sitting in his lap while he read, and how comfortable, safe, and cozy she had felt sitting on her father's lap, and the sense that he was always there for her. In truth, my sister adored her father and always will.

There are so many childhood memories of dad's joy. I remember a game we used to play where I stood on dad's shoes while facing him as he walked. I remember him beating me at tennis as teenager and he danced across the court in glee saying, "Game. Game. Game. Game. Game."

My dad took great joy in my achievements. When he got home from work one day, he proceeded upstairs when I yelled to him that I had gotten into Bates College. He came running down the stairs with a big smile and wanted to hear all about it. That same sense of joy and smile has been repeated many times at graduations, ordination, and installations.

I was interested last week when my brother noted that my father did not consider himself a musical person and that is quite true. However, I will always remember his humming and whistling. And he took great joy in the bagpipes. As a teenager, I went with him to take Cousin John Grant to work: playing the bagpipes outside a Halifax department store. He would hum "Scotland the Brave." He would hum his favorite hymns as we came home from church. There were two favorites that he sang in the car on our summer trips. (Singing) "My bonnie lies over the ocean. My bonnie lies over the sea. My bonnie lies over the ocean. O, bring back my bonnie to me, to me. O bring back. O bring back, O bring back my Bonnie to me, to me."

Or…

> (Singing) "Down by the old, not the new but the old, mill stream, not the river but the stream. Where I first, not last but first, met you, not me but you. You were sixteen, not seventeen, my high school queen, not king but queen. Down by the old, not the new but the old, mill stream, not the river but the stream." Dad particularly relished the echo.

I didn't really have a great sense of what dad did when I was a child. I knew he was a doctor and a geneticist and distinguished in his field, but I don't think I had the sense of his greatness and his contributions until I was in college. I recall a discussion in high school with friends talking about what their parents did for a living. My friend Gordon (who is here today), when I started to speak, rebuked me by saying, "Shut up, Victor, your father invented DNA!"

My brother and I as kids went on a very memorable trip with my dad on Amtrak to New York and stayed at the Americana Hotel. We never left the hotel. My brother and I stayed in the room and ordered room service. It isn't like we saw the city or took in a Broadway show. For me, it was just great to be traveling with my dad.

One Washington's birthday, I traveled with dad and one of his Fellows to Amish Country in Lancaster County, Pennsylvania. I think what was special about that day is he set a personal best for farm visits that day. I believe we made 23, but I could be mistaken. What was memorable for me was meeting Amos Stoltzfus, a fifth grader with a sixth finger on one of his hands. Amos had a very special challenge in learning how to write. Later, I've learned that the sixth finger is a symptom of McKusick-type achondroplasia. Don't ask me to say that again. Amos was a student in the one-room schoolhouse of Sarah Fisher, an Amish teacher who was one of dad's contacts and hosts in the Amish community. I was always impressed with how dad related with his Amish patients; they seemed to admire him and get such joy at of his visits.

I noticed one of my dad's little professional quirks. When in book stores, he would take time to sign his own books. He did this not as part of a book signing mind you but just slip a couple of his books from the shelf and sign, "Victor A. McKusick." I think he liked the idea of the new owner being surprised and happy as the reader found a signed copy of MIM, or "Green Genes" which was dad's playful nickname for an early genetics publication.

I never really understood what *Aequanimitas* meant until this last year of dad's life. From what I understand of William Osler, dad's idol, the great physician spoke of *Aequanimitas* as a balance—a balance between engagement with the patient or student and yet a professional detachment of expertise. When I read the 2000 interview with dad about his life's work, I figured out what it meant: it is not only something he lived, it is not only what many of us seek in life, but I can understand it as a goal that I seek as a pastor. In 1957, A McGee Harvey named dad as Director of the Moore Clinic, a clinic that had studied syphilis. Because of Dr. Moore's fundraising, grants, and various endowments, dad and his enterprise which laid ahead was independently funded. He had the opportunity to follow his intellectual curiosity and did. Furthermore, as the story goes, there were many naysayers of the genetic endeavor, yet dad stuck to the course of his life. He went about it with *Aequanimitas.* Why then is *Aequanimitas* so important for us all? Despite all the distractions, despite all of our desires to be reactive, to carry the day, when we can stay on course and follow that balance, follow that center, we are at our most meaningful. Last night, my cousin Paul mentioned that the necessity for equanimity is really present in many professions. I know as a pastor; I find the most satisfaction and meaning when I am visiting someone or in a meeting, and I am not only aware of myself and but engaged with those around me. I think this is *Aequanimitas.*

The last Saturday of his life I said goodbye to dad for the last time. "Dad I love you, and thank you for everything that you have done for me" and then kissed him on his forehead. I was feeling woefully inadequate for not having had long and deep, meaningful talks with him in his last months of life. Then Lori reminded me that my dad had read an article by a writer in Central New York. The author had scribed an article about my dad and me for Father's Day in a local paper. Somehow, that would always have been a better medium for dad to hear what I thought and felt about him, rather than any long heartfelt talk. So Good-bye dad. I love you!

Appendix B 3. Eulogy by Clair A. Francomano

Eulogy for Victor A. McKusick, M.D.
Teacher, Mentor, and Role Model

Clair A. Francomano

We are here to commemorate, and to celebrate, a remarkable life. Others have discussed Dr. Victor McKusick's prodigious contributions to medicine, to the field of medical genetics, and to Johns Hopkins. My task, as he requested, is to discuss Dr. Victor McKusick's legacy as a teacher, mentor, and role model.

I first met Dr. Victor McKusick when I was a high school student participating in the Jackson Laboratory summer program in Bar Harbor, Maine. During that heady summer of first exposure to the principles of genetics and scientific research, we students had the privilege of attending the Short Course on Medical Genetics, which Dr. McKusick had cofounded with Dr. Thomas Roderick of the Jackson Laboratory in 1960. There was a cocktail party during the course, and I was selected, by virtue of my legible penmanship, to write the nametags for attendees. A secondary assignment was to serve hors d'ourvres. And so it was, in 1971, that a high school genetics neophyte from Yonkers, New York, first encountered the Drs. McKusick face to face, on the lawn of the Drs. Earl and Margaret Green overlooking beautiful Frenchman's Bay in Bar Harbor, Maine.

Eventually, I came to know him as a medical student, a house officer on the Osler medical service, a genetics fellow, and a faculty member at Johns Hopkins. He sent me to Lancaster County, Pennsylvania, and Holmes County, Ohio, to research the Amish. He introduced me to the little people and the tall people with Marfan syndrome and their support groups, and to dozens of colleagues from around the world. He was unstintingly generous of his time and whatever resources he had at his disposal. He encouraged my research and supported my career in every imaginable way. I know that he did this for every one of the many who came before me. There are 121 people on the list of genetics trainees supervised by Victor A. McKusick, M.D. We have all been gifted by his enthusiasm, his innate curiosity, his life, and his work. By his example and the opportunities he made available to us, he taught us to explore deeply, to think critically, and to write clearly. He molded us into the professionals we became.

As I thought about the fundamental lessons that Dr. Victor McKusick left us as a teacher, mentor, and role model, three essential teachings emerged:

1. Work very hard
2. Always stay positive
3. Share what you know

I would like to take a few minutes to elaborate on each of these lessons. I should say that he never would have told us directly to do any of these things—his teaching was always by example rather than by exhortation.

1. *Work very hard.* Truly, Victor was always working. He just kept at it, day in and day out. An avowed Lark (as opposed to a night owl), he would awaken early in the morning and start working at 5 A.M. He worked on the way to the hospital when Anne would drive him in the car. He worked on the weekends. He worked while he was away and he worked at home. He was never not working. When I was a fellow, we were making rounds one Monday, and he said to me, "I wrote a book this weekend." Just like that. He accomplished great things, in part, because he worked very, very hard, all the time. It gave him his greatest joy and pleasure.

   He expected hard work from his students and trainees, but never more than he gave himself. Dr. Judith Hall told me a story of revising her manuscript on Thrombocytopenia-Absent Radius Syndrome when she was working with him as a fellow. She was on the 13th draft and really slogging it out, and she asked Dr. McKusick, "Is it always this hard?" With his characteristic brevity, he replied, "Yes." Nonetheless, it was exhilarating to share in his passion, his enthusiasm, and his commitment.

   Dr. Michael Bliss, in his essay about Sir William Osler on the 150th anniversary of his birth [1], wrote: "No one has ever loved the medical life more than William Osler did. From the day he decided to become a doctor until the day of his death he lived, breathed, ate, slept, talked and wrote medicine." Dr. Victor McKusick, who was an avid Osler historian, followed closely in this path. The morning of the day he died, he watched the streaming video from his beloved Short Course in Bar Harbor.
2. *Always stay positive.* I have known Dr. Victor McKusick for 37 years, and I worked very closely with him for over 20 of those years. In all that time, I never heard him complain or say anything negative about a situation or another person. I do not believe that I know another person about whom I could say the same thing. He was truly remarkable in that way. If he saw a situation that was not to his liking, he worked to change it, quietly, without rancor, and without discord. I never heard him raise his voice.

   Again, I would like to quote Dr. Bliss [1], who wrote about Osler: "Someone has said that the real trouble with Osler as a role model is that he never had a bad day, and no one can come close to matching him." I believe the same could be said for Victor McKusick. He epitomized Osler's principle of aequanimitas.
3. *Share what you know.* Surely this is his greatest legacy. All of us who trained with him benefited from this essential aspect of his being. It was simply a part of who he was. He took legions of house officers up to the Hopkins Dome to share his knowledge of Hopkins history. Mendelian Inheritance in Man, now OMIM, was his way of making the entire evolving knowledge base of human genetics available to the widest possible audience. Not only did he share what he knew, freely and without reservation, he brought people together to share what they knew with each other. The Birth Defects meetings and Short Stature Symposia at Johns Hopkins, the Bar Harbor and Italian genetics courses, and even the Human Genome Organization were born of his innate understanding that for a body of

knowledge to survive and prosper, it must be shared, discussed, written about, and preserved.

If I may be allowed one final quote from Dr. Bliss concerning Sir William Osler [1], it is this: "Few of us are as fortunate or as gifted in our strivings to take the good of every hour and contribute to human betterment." Certainly, Victor McKusick was one of the very few.

I had the opportunity to enjoy Anne and Victor's hospitality on many occasions.

Together they were two parts of a grander whole. I would like to take this opportunity to thank Dr. Anne McKusick for sharing him with us, and especially for her kindness, hospitality, and warm friendship these many years.

The last time I saw him, 1 week before his death, I asked Victor if he had any thoughts about the afterlife. He smiled and said, "I'm keeping an open mind." This was so typical of him. Wherever he is now, I am sure he is taking notes in one of those little brown books.

He has been an integral part of my entire professional life, as he has for so many others. For those of us fortunate enough to be able to say that, it is difficult to imagine a world without Victor A. McKusick. Perhaps the best way we can honor his memory is to perpetuate that which he taught us, and lived by: Work very hard, always stay positive, and share what we know, widely and deeply, with whoever will stop to listen and learn.

Although these simple words cannot begin to express the magnitude of our debt to you, Thank you, Victor, from all of us.

Clair A. Francomano, M.D.
August 2, 2008

Bibliography

1. Bliss M (1999) Osler at 150. Canad Med Assoc J 161:831–34

Appendix B 4. Eulogy by Stephen C. Achuff

Eulogy for Victor Almon McKusick
Stephen C. Achuff

It was my great good fortune to be Chief Resident, along with Tom Inui, during Victor's first year as Chairman of Medicine 1973–1974. We did not really know him before that, other than by reputation, and he certainly did not know us, as we had been foisted on him by his predecessor, A. McGehee Harvey. Frankly we thought he must have been a little rusty as a clinician having spent the previous 20 years or so studying all those strange and uncommon genetic disorders. Being perhaps a bit cocky, we saw it as our duty to bring him back into mainstream medicine—and, we had a little help from Philip Tumulty, who was probably the master clinician and teacher of that era. Every weekday morning, the four of us would meet in Victor's office on Blalock 10 and discuss the patients who had been admitted the previous day. Needless to say, Victor was a quick study, and when something came up that was unfamiliar or that piqued his insatiable curiosity, he would pull out one of his little brown books and make notes to himself for later reference or research. It was in those informal meetings and bull sessions that usually followed, that the concept of the Firm System was born—one of Victor's most enduring innovations and of which he was justifiably proud. The Housestaff had grown larger over the years to the point of not only being unwieldy, but something of the sense of camaraderie and family that Victor remembered and cherished from his own days as an intern and resident were in danger of being lost. Plus, medicine on a larger scale had changed what with a new focus on intensive care, shortened hospital stays, governmental regulation, and so forth. So over the next couple years, Victor reorganized the Medical Service—retaining what was good about traditional Osler training and also adapting to the exigencies of medicine in the 1970s. As we came to know Victor on a personal level, it was nothing short of mind-boggling to witness firsthand his encyclopedic knowledge, his phenomenal powers of recall, and his prodigious work habits. This last was brought home early in the year when his secretary called me over to her desk one morning as she was transcribing his letters and papers and had me listen to the cassette—In the background behind his dictation was the unmistakable twittering of song birds—as usual, he was up before dawn accomplishing more before the sun rose than most of us do in a full day—something undoubtedly ingrained during his youth growing up on a dairy farm in Maine.

And then there was Victor's passion for history and his commitment to remind us all of the great heritage of the department and the responsibility we shared to continue and promote that heritage. As best I can tell, he never really explained why he took students and residents to the top of the Dome, talked more knowledgeably than anyone about the great Hopkins physicians of the past, and showed the room where Osler wrote the first edition of his Principles and Practice of Medicine in 1892. He did not force this on us, he just did it, he obviously loved doing it, and nobody could do it better. One tangible sign that will live on as long as there is a Department of

This photo was taken in the room where Osler wrote the first edition of his textbook, "Principles and Practice of Medicine", in 1892. Dr. McKusick is sitting in a high-backed desk chair that was Osler's. Dr. Stephen Achuff is sitting on Osler's desk and in the background is a photo of Osler writing the textbook. Both are wearing the Osler tie—"Aequanimitas" which was created by Dr. McKusick in 1977–78. Everyone who trained on the Osler Medical Service is obliged to wear the tie on Fridays (Courtesy of Stephen C. Achuff and Andrew Harrison [Alan Mason Chesney Archives of the Johns Hopkins Medical Institutions])

Medicine at Johns Hopkins is the Osler tie for men and scarf for women with the word Aequanimitas on a shield—the idea having been suggested by one of his residents, John Beary, after he had spent an elective at the Radcliffe Infirmary in Oxford. Victor, ever the incurable Anglophile, leapt at the idea and started a tradition that has been emulated by virtually all the other departments at Hopkins as well as many around the country. Incidentally, one of the picture boards here today (in the narthex) is a photo of a very contented Victor wearing his Osler tie, sitting in Osler's chair (appropriately), in the Osler Textbook Room.

Anne called about 3 weeks ago and said Victor wanted to see me. When I arrived at their apartment at Blakehurst and sat down next to him, I was a little taken aback by his first words—not so much because he asked me to speak at his funeral, that would be a great honor and I told him how much I appreciated it, but more because of what he wanted me to say—and why. I am paraphrasing, but essentially he said that for a long time he had been disappointed that when people talked or wrote about him, the focus was always on his scientific achievements and what was given short shrift, if any mention at all, were his contributions to education of medical students and young physicians, to preservation of the very best traditions of Hopkins and Oslerian medicine—in his mind, and I think rightfully so, some of his most important and enduring achievements. In point of fact, he had led for more than a decade the largest department in Johns Hopkins University and been Physician-in-Chief at arguably the number 1 hospital in the country. Victor was a modest man—who bore fame and success with uncommon humility, but clearly he took great satisfaction in having been William Osler's successor and having preserved the best aspects of Oslerian medicine. I think he was undeniably justified in that accomplishment, and I hope I have been able to express some of that this afternoon.

Appendix B 5. Eulogy by David L. Rimoin

Eulogy for Victor A. McKusick
David L. Rimoin

Victor McKusick is a legend among his peers, respected and revered as the "Father of Genetic Medicine," as well as a master clinician, scientist, medical historian, writer, teacher, and mentor.

Victor wanted me to speak on the establishment of medical genetics as a medical specialty and the vision that he and many of his students have been working toward in the field.

Victor was a physician, first and foremost, and followed the Oslerian tradition of paying close attention to his patients, seeking insight into the cause and management of their disease and problems.

He had become interested in genetics as an undergraduate at Tufts. He also experienced applied biology on the dairy farm as a child, and he wrote a paper with his older brother on coat color inheritance in Jersey cattle, which was submitted but never published.

His first genetic paper was in1949 which dealt with the Peutz-Jeghers syndrome, characterized by skin pigmentation and intestinal polyps.

He claimed that he then got training in cardiology because there was no such thing as medical genetics. With apologies to Dr. Ross, he said that he found genetics much more challenging, exciting, and novel.

Through his cardiology interests, he came across families with Marfan syndrome and aortic aneurysms and conceived that this was a syndrome in which one could think of the many effects of a single gene that affected one element of connective tissue, wherever it was in the body.

He became interested in the clinical and genetic aspects of the various heritable disorders of connective tissue—a term and concept that he introduced. He published his first book, Heritable Disorders of Connective Tissue, in 1956.

Over the years, he excelled in the clinical nosology of these and many other genetic disorders. Some of his colleagues viewed his research as the medical equivalent of stamp collecting and some wondered if it was even science. But Victor perceived that the future of medicine—and insight into the molecular gears and switches that are the science of life—lay in the direction he was heading. If getting there required going house to house, examining babies, asking about grandparents, etc., he was more than happy to do it. He claimed that he had a great deal of fun tracing hemophilia in early New England families. In jest, he said that we would knock on doors and say, "Take me to your bleeder."

Victor would travel far and wide to see unusual patients that could teach him about genetic diseases. When I was a fellow, Victor taught me the power of clinical research and the value of studying rare diseases to prove basic biologic principles. I accompanied him to meetings of the Little People of America and even the so-called freak show at the Barnum and Bailey Circus at Madison Square Garden to find new research subjects.

Victor did not limit himself to descriptive studies. Enlisting the help of many basic science colleagues, they were able to define the biochemical and molecular defects in many disorders. He was especially effective in crossing basic science and clinical lines and across specialties.

He recently reached his ultimate goal in Marfan syndrome with over 50 years of work, starting with description and natural history, to the exciting work of Hal Dietz, the first McKusick professor, of finding a potential pharmacologic therapy for the aortic disease that first attracted his attention. This is a classic example of clinical genetics at its best.

In 1957, Victor took over the Moore Clinic and started a division of medical genetics within the Department of Medicine. As a general hospital division, this brought genetics into the mainstream of clinical medicine.

Victor built a division of medical genetics with cytogenetics, biochemical genetics, population genetics, and immunogenetics components. His cytogenetic lab was perhaps the earliest clinical cytogenetics laboratory in any large general hospital in this country. Pharmacogenetics also got started in the Moore Clinic with the study of the genetics of metabolism of the antituberculosis drug, INH.

He recruited several hundred fellows in medical genetics from around the world over the years, many of whom became international leaders in the field—he was a true pied piper. A couple of years ago, Victor was in LA and he attended the large Sports Spectacular fundraiser for genetics at which the legendary coach John Wooden was being honored. I introduced Victor to this audience of 1,800 sports fans as the John Wooden of Medical Genetics—Victor really liked that comparison!

Victor started a monthly journal club at his home where each fellow had to review six to seven medical journals and bring in a short description of any new papers on genetic disease. He began publishing an annual review of new findings in medical genetics, which evolved into his classic Mendelian Inheritance in Man, long considered the bible of medical genetics—a constantly updated catalog of genetic diseases and the genes that caused them. This was before the day of the word processor. He put it on the computer in 1964 and then produced a printout book that became one of the first computer-generated medical texts.

His involvement with the Amish was a fringe benefit of university committee work with the Johns Hopkins Press, when in the fall of 1962 he reviewed the manuscript for John Hostetler's book called Amish Society. His studies of inherited disorders in the Amish uncovered about a dozen previously unrecognized, inherited conditions and served as a model for studies in similarly isolated populations elsewhere.

His involvement with the skeletal dysplasias was an outgrowth of the Amish studies. He was first invited to the National Convention of Little People of America in 1965 as a result of a story about their studies in dwarfism in the Amish in Time Magazine. He was later named an honorary life member of LPA. He was also an honorary fellow of the American Academy of Orthopedic Surgeons and claimed that he was the only member of those two organizations who was more than 6 ft tall and did not operate.

The pioneering observations that G6PD and color vision are closely linked on the X chromosome and that the Duffy blood group is located on chromosome no. 1 came from his unit. Victor was clearly a clairvoyant. At the birth defects meeting in 1969 in The Hague, he suggested that mapping the human genome would be the way to go to unravel the mysteries about lots of birth defects and genetic diseases.

In 1973, together with Frank Ruddle, he began the international workshops on gene mapping in man and pioneered the use of computers for linkage work. He approached the genome as an organ and began writing about the morbid and functional anatomy of the human genome.

The human genome project was strenuously debated because many people thought it was a crazy and very expensive idea, that it really was not science but just brick counting.

As a leading proponent of completely mapping the human genome, Victor served as an enthusiastic advocate, advisor, and kibitzer of the Human Genome Project and served as the founding president of the International Human Genome Organization (HUGO).

Thus, Victor was a true cartographer, whose efforts not only led to the mapping of the human genome, but he also put genetics on the medical map.

Through his fellowship, courses, meetings, and writings, Victor was an extremely effective teacher and salesman of medical genetics.

His annual genetics course at Bar Harbor, Maine, is widely credited with training generations of genetic medicine practitioners and scholars. This course was first given in 1960, and last week, Victor attended the course by computer streaming even to his last day.

This long-standing course attracted a large number of health science faculty members, including a fair collection of the department chairs and even deans. One of the major objectives of the course was to upgrade the teaching of genetics by teaching the teachers. Indeed, a large proportion of people in genetics in this country have attended the course at some stage of their careers. The multiplier effect of this "teaching of teachers" was a catalyst in the development of medical genetics as a specialty.

Victor also created the annual Clinical Delineation of Birth Defects meetings—a yearly meeting dealing with nosology in clinical genetics.

Although at first Victor thought there was no need for a board in medical genetics, he changed his mind and joined us in creating the independent American Board of Medical Genetics. A decade later, the ABMS admitted the ABMG as the first new fully independent medical board in 20 years, and as a full specialty.

We then formed the American College of Medical Genetics that became the 24th recognized medical specialty by the Council of Medical Specialties. We instituted an annual meeting and combined it with the ongoing March of Dimes (MOD) Meeting, which were originally Victor's Clinical Delineation meetings.

Victor was always one to two decades ahead of his time: creating a division of medical genetics, inventing spectophonocardiography, predicting the usefulness of linkage and gene mapping, pioneering the use of computers in publishing, and medical use of the internet.

He made major contributions to biochemical, molecular, and population genetics without being in a lab or getting formal training in mathematical genetics or computer technology. He considered his lab to be the clinic and the library.

Thus, enormous energy, a rigorous work ethic, effective time management, the ability to foresee the use of multiple new technologies, rapid response to new findings, coupled with the warm personality of kind and sympathetic doctor, teacher, and friend made for a remarkable individual, who has inspired thousands of students, doctors, and scientists around the globe. Almost every physician in the future will benefit directly or indirectly from his research, teaching, and his persistent and successful championing of the human genome project.

I have been a McKusick disciple for 45 years—he was my teacher, my mentor, my role model, and my champion and was much like a father to me and an uncle to my children. I shall miss him dearly.

Appendix B 6. Eulogy by Richard S. Ross

August 2, 2008
Richard Starr Ross

I have seen Victor several times since he returned from his trip to receive the Japan prize, and I have never seen him happier. He knew better than the average patient what the future held, but he was happy. He was enthusiastic in describing the presentation ceremony and his conversation with the Emperor and the Empress of Japan. Nova Scotia and Longfellow's poem, Evangeline, were the subject of discussion with the Empress.

You can sense Victor's enthusiasm in the photograph taken on that occasion. He was proud and felt that the awarding of the prize represented recognition of his lifetime work as a medical scholar. His pride was obvious and justified by all his accomplishments. We can all be thankful that his health permitted him to attend this ceremony.

I was somewhat startled but greatly honored a few weeks ago when Victor asked me to speak at his funeral. I was selected as someone who had known him for over 60 years and watched him with awe from numerous vantage points.

I first met Victor in 1947 when I came to Hopkins as a lowly intern, and he was a lofty assistant resident. During my first few years at Hopkins, Victor took me under his wing and taught me many things, among which was how to write medical papers. Victor was an absolutely superb writer and editor.

> [I even considered the possibility that he asked me to speak at his funeral several weeks in advance so that he could edit this talk. I am sure he could have improved it, but I didn't ask.]

The secret to his success as a writer was his ability to manage his time and concentrate on the task at hand. He also was blessed with superior intelligence, an awesome memory, a dictionary-size vocabulary, a boundless energy, and the ability to work early in the morning.

We wrote a paper together that was published in 1953. This paper is number 1 on my list of publications, but it is number 26 on Victor's list. This gives you some indication of how far ahead of the pack he was at the very beginning of his career.

In 1975, when he was considered for the post of physician in chief and chairman of the Department of Medicine, there was a concern that he would not be a good choice because he had moved so far into research on rare, exotic diseases. Questions were raised as to whether he would be able to run a department responsible for training young people in the broad field of internal medicine. These doubters were wrong. He quickly cleaned the rust off the skills acquired as an Osler chief resident and became a superb department leader. The interns and residents stood in awe of his encyclopedic knowledge and loved him as a person. Steve Achuff will speak about this period in more detail.

[In the last few days, I encountered a young woman, at least she was young 60 years ago, who had worked for me and Victor as a technician. She had made charts and graphs and slides for us. She was reflecting on Victor's death and said that she remembered that he always stood up when she entered his office to show him her work and only sat down after she sat down. I timidly asked her if I stood up when she came to see me and she replied, "I can't remember."]

In 1984, Victor was 63 years of age and 2 years from retirement, and he was having trouble filling some of the key positions in his department. Victor did not want to step aside early, but he could see that it would be best for Hopkins if he made it possible for his successor to fill the open slots. Another factor was his recognition that the field of medical genetics was about to explode, and he wanted to be out in front.

I was the dean at this time and have reviewed an exchange of handwritten letters I had with him in 1984 in which the pros and cons were laid out, and it was decided that it would be better for Hopkins and better for him if he gave up the chair of medicine and moved into genetics.

He left the position of physician in chief in 1985 while he was ahead and led Hopkins to the lead position in medical genetics. This is the most dramatic example of Victor's superb timing in switching fields just at the right time.

Friends were very important to Victor. He and nine friends from the Osler house staff of the 1950s and wives met one Saturday night a month for what was called "supper club." This organization still exists, although diminished in numbers and frequency of meeting, was always very important to Anne and Victor. Victor relaxed and enjoyed the company of old friends. He appreciated a good joke and had a strong, infectious laugh.

No tribute to Victor can be completed without recognition of Anne, in the background when appropriate, but able to move in decisively when needed. On one of my recent visits to see him, he was ruminating about what his life would have been like if he hadn't come to Hopkins, and first on his list was, "I would not have met Anne." I can add to that by noting that he would not have been so successful without Anne.

I am wearing the "Osler necktie" designed by Victor and to be discussed later by Steve Achuff. The word on the necktie, Aequanimitas, was used by Dr. William Osler, later Sir William, the first physician and chief at Johns Hopkins, to describe the inner calm that all good physicians must have. Victor was a great admirer of Osler and patterned his own career after this great man of medicine. He was also very fond of the word Aequanimitas and what it stood for.

At this very moment, I can picture Victor and Sir William together, and Osler says, "welcome McKusick, my worthy successor, will you join me on rounds?" They walk together into the ward.

A final word to you, Victor, Aequanimitas forevermore.

Appendix C
Obituary from Francis Collins and Obituary from Aravinda Chakravarti

Reprinted From *Science, 321*, 925 (2008)

Victor A. McKusick (1921–2008)
Francis S. Collins, National Institutes of Health, Bethesda, MD 20892, USA.

It is the rare scientist who is universally recognized as the founder of a field. Even rarer is the one who witnesses his vision evolve from a solitary pursuit into a major discipline. But such was the life of the father of medical genetics, Victor Almon McKusick, who died on July 22 after more than a half-century of pioneering research, mentorship, and leadership.

McKusick was the driving force for moving genetics beyond the tidy realm of flies and mice in the research lab into the messier realm of the medical clinic. In 1957, he established a medical genetics clinic at Johns Hopkins, the model for what would become more than 100 such clinics nationwide, and unleashed a flood of discoveries that demonstrated connections between genes and diseases. Today, thanks largely to McKusick's creative and tireless contributions, the field of medical genetics is at the center of medicine.

While the influence of some scientific leaders wanes with their passing, McKusick's will only grow with time. His work created a dynamic legacy that lives on in the thousands of physician researchers he mentored and the marvelous, updatable resources he created. Nearly, every medical geneticist, including myself, can trace his or her scientific lineage back to McKusick.

The first of my many sips from McKusick's seemingly bottomless well of knowledge came in 1981. As a medical genetics fellow at Yale, I was confronted with a puzzling infant with a congenital intestinal obstruction. The child had a sibling with an identical condition. I was stumped about the diagnosis, as was my attending physician, Uta Francke. So, we consulted McKusick's landmark text, *Mendelian Inheritance in Man*, and came upon entry "*243600, familial apple peel jejunal atresia," a perfect description of the child's condition. Whereupon, Uta said, "This is wonderful. Victor should win the Nobel Prize!"

K.R. Dronamraju and C.A. Francomano (eds.), *Victor McKusick and the History of Medical Genetics*, DOI 10.1007/978-1-4614-1677-7,

Scientific prizes probably were not on anyone's mind when McKusick and his identical twin, Vincent, were born on October 21, 1921 on a dairy farm in Parkman, Maine. In fact, McKusick considered becoming a minister until age 15, when he developed a serious streptococcal infection and came away inspired by the physicians who saved his life. After spending 3 years as an undergraduate at Tufts, McKusick enrolled in 1943 in the Johns Hopkins School of Medicine, from which he graduated in 1946 and where he held a series of distinguished professorships up until his death.

While he authored many seminal papers and books on genetics, McKusick also delighted in sharing the story of his first encounter with the world of scientific publishing. As a medical student, he and his brother Robert (who became a dairy farmer) wrote a paper on the inheritance pattern of coat color in Jersey cattle, describing how the apparently dominant fawn color was actually a recessive trait. They eagerly sent their genetics paper off to the *Journal of Heredity* but never got a response. Fortunately, both for McKusick and the scientific community, future journal editors would display better judgment.

Among his key publications was a 1966 paper describing the first mapping of a human autosomal gene, the Duffy blood group locus. He also clinically defined a series of connective tissue disorders, the most famous of which is the Marfan syndrome. The tour de force of McKusick's publications, however, remains his constantly updated *Mendelian Inheritance in Man*, a catalog of human genes and genetic disorders. This classic reference first appeared in print in 1966 and contained about 1,500 entries. Today, an electronic version features nearly 19,000 entries.

No description of McKusick's impact on science is complete without mentioning his prescient call for mapping the human genome. In August 1969, at the International

Conference on Birth Defects in The Hague, McKusick proposed that mapping all human genes would be useful for understanding basic derangements in birth defects. "The proposal reflected the exuberant mindset that followed the first Moon landing by Apollo 11," he recalled in a 2006 article. But the idea met with perplexed silence, in part because no one—not even McKusick—was clear on what methods could be used to achieve such an ambitious goal. Thankfully, McKusick was not one to give up easily and was a strong advocate for the Human Genome Project, stepping forward in the mid-1980s to serve as the founding president of the Human Genome Organization.

During a career that spanned an impressive 60 years, McKusick received many accolades, including election to the National Academy of Sciences in 1973, the Albert Lasker Award for Special Achievement in Medical Science in 1997, the National Medal of Science in 2001, and the Japan Prize in Medical Genetics and Genomics in 2008.

Medals and proclamations aside, I suspect that one achievement that McKusick would most want to be remembered for is his role in establishing the legendary "Short Course in Medical and Experimental Mammalian Genetics." This 2-week event, held each summer at The Jackson Laboratory in Bar Harbor, Maine, has had a profound influence on medical genetics. Since it began in 1960, more than 5,000 clinical specialists, educators, and others have had the privilege of learning about the latest advances from some of the best minds in the field. In fact, according to Anne, McKusick's physician wife for 59 years, hours before her husband died peacefully of cancer at their home outside Baltimore, Maryland, he had enjoyed watching the live streaming video of this summer's "Short Course."

More than a century ago, the father of modern medicine, Sir William Osler, wrote, "To wrest from nature the secrets which have perplexed philosophers in all ages, to track to their sources the causes of disease, to correlate the vast stores of knowledge, that they may be quickly available for the prevention and cure of disease—these are our ambitions." Victor McKusick, who appropriately held the Osler Professorship for many years at Johns Hopkins, lived that vision better than any other physician of the last half-century.

Obituary

Victor Almon McKusick (1921–2008)

Quiet Revolutionary in Genetic Medicine

Reprinted by permission from Macmillan Publishers Ltd: Nature, 455, 46 (2008).

Following the complete sequencing of the human genome, we stand at the beginning of an era that promises medical treatments tailored to an individual's genetic makeup. No one is more responsible for this revolution than Victor McKusick, who died on July 22. McKusick was the first to understand that systematically mapping human genes predisposing the bearer to disease, which many considered no better than stamp collecting, was a route to a new medicine. In this and other ways, he was instrumental in molding the discipline that we now call genetic medicine and in making genetics the basic science of medicine.

McKusick was born on a dairy farm in Maine in 1921. His early ambition was to enter the ministry. At the age of 15, however, a streptococcal infection of his arm that required a long hospital stay and treatment with one of the first antibiotics made him rethink his future. His identical twin Vincent chose to study law. Victor, by contrast, after initial education at Tufts University, entered Johns Hopkins University in 1943 to pursue medical training, making a name for himself in cardiology.

Like medicine, genetics came to him by chance. His fascination with one teenage patient who suffered from intestinal polyps and melanin spots, and later with three members of a family who exhibited the same syndrome, provided him with firsthand experience of the basic principles of genetics. One was the need to recognize patterns of inheritance, in this case, dominant as opposed to recessive, that suggested mutations at one genetic location. Another was the need to distinguish between mechanisms: in these patients, were two genes involved, one for polyps and one for spots, which were coinherited (linkage), or were polyps and spots different manifestations of the same gene (pleiotropy)? McKusick was thus well armed when he subsequently came across patients with Marfan syndrome—with its dominant inheritance and remarkable pleiotropy affecting the aorta, eye, and skeleton—which, he argued, arose from mutations in a single gene.

Similar patients and their families were to prove pivotal in his conversion to genetics, which was completed by 1957. Asked to direct a chronic-disease clinic by his boss, McKusick argued that "genetic disease is the ultimate chronic disease, since it's lifelong," and seized the opportunity to reshape the Moore Clinic at Johns Hopkins to create the first unit devoted to medical genetics. He learned his trade by doing: by using the rudimentary cytogenetic, biochemical, and population (quantitative) genetic methods then available. He soon became convinced of the three guiding principles: the value of knowing a gene's location in the human genome, the value of accumulated genetic information, and the value of disseminating this new information widely and rapidly.

Given the individual rarity of most hereditary disorders, McKusick knew that he had to learn about the experiences of others and to share his own. He was a prolific organizer, of both ideas and facts, a trait most notably made manifest in 1966 in *Mendelian Inheritance in Man (MIM)*, the first edition of his catalog of all known genes and genetic disorders. The final print edition appeared in 1998, but since 1987, it has also been available as full-text online, with a free database (www.ncbi.nlm.nih.gov/Omim). It now has some 19,000 entries, with more than 70% of the content having been produced by McKusick himself. This is his most lasting achievement—it is a deep resource and knowledge base, without which clinicians and any manner of biologist would be intellectually orphaned.

One of McKusick's preoccupations was with cataloguing the location of each human gene associated with a disease and thus to create a disease map of the human genome. He did this not only through his own pioneering studies but by beginning—chiefly with Frank Ruddle—a series of human gene mapping workshops. Subsequently, he was an influential voice in organizing the international community around the Human Genome Organization (HUGO, fondly called Victor's HUGO). For him, the *raison d'être* of mapping, which he articulated in 1969, well before anyone understood or believed it, was that mapping all human genes was the best way to understand the basic malfunctions causing birth defects.

The existence of *MIM*, together with McKusick's mapping preoccupation, were the two most persuasive factors in favor of the public project to sequence the human genome. McKusick himself was on the US National Research Council committee that recommended the project and was one of its prime cheer leaders. He was among those who argued for a "map first, sequence later" approach and was a supporter of mapping and sequencing other species and of tackling the whole genome rather than only the known functional genome.

As a pragmatist, however, McKusick was also attracted to Craig Venter's idea of sequencing expressed sequence tags (nucleic-acid snippets that encode only a portion of functional genes). He supported both the public sequencing project and Venter's private sequencing effort at Celera (he was a trustee of Venter's eponymous institute) because he believed that the genome could thus be completed sooner. The leaders of both the public and private sequencing ventures (Francis Collins and Venter, respectively) paid their respects at his funeral service.

McKusick made a research on the human species, despite its poor genetic properties of few offspring and long generation times, a treasure trove for uncovering new genetic mechanisms. He also provided a glimpse of the future for genetic medicine in an interview given in 2001: "I think the medical geneticist will spend much more time overseeing gene screens, or genome screens, interpreting the results to individuals, and designing programs to make the most of the strong points of the genome and to avoid troubles from some of the weak points in the genome." Spreading the word was a vital part of his legacy—as, for example, in the influential "Short Course" in mammalian genetics, held annually at the Jackson Laboratory in Bar Harbor, Maine, which he founded in 1960 and codirected.

In the long journey to his many accomplishments, Victor McKusick was accompanied by his rheumatologist wife Anne. Those accomplishments are all the more remarkable for having been achieved without his once raising his voice. But then, a man who had genetic institutes named after him in Baltimore, Bologna, and Beijing had no need to draw attention to himself.

Aravinda Chakravarti
Aravinda Chakravarti is at the McKusick-Nathans Institute of Genetic Medicine, Johns Hopkins University, Baltimore, Maryland 21093, USA.
e-mail: aravinda@jhmi.edu

Index

K.R. Dronamraju and C.A. Francomano (eds.), *Victor McKusick and the History of Medical Genetics*, DOI 10.1007/978-1-4614-1677-7,

Printed by Publishers' Graphics LLC